Prejudice Against Nature
A Guidebook for the Liberation of Self and Planet

by

Michael J. Cohen, Ed.D.

Director, National Audubon Society Expedition Institute
Adjunct Faculty, Lesley College Graduate School

with an Afterword by Jim Swan, Ph.D.
co-author of

Environmental Education: Strategies Toward a More Liveable Future

COBBLESMITH
Box 191 RFD#1
Freeport, Maine 04032

To Frank Trocco and Dan Tishman

In fond memory of Ben Fielding

The author gratefully acknowledges that many special people and places knowledgeably or inadvertently became a vibrant part of myself and this book. To all my students and sponsors, I extend my heartfelt appreciation. See Appendix E.

Now in their 15th year of operation (1983), the experiential learning courses and degree programs of the National Audubon Society Expedition Institute are an innovative, accredited approach to environmental education. They are offered under the auspices of the National Audubon Society, Lesley College, the University of the State of New York, and the Institute for Expedition Education.

The ideas and outlooks conveyed by this book are the collective and individual thoughts of Expedition Education participants; they do not necessarily represent the views of the Institute's aforementioned sponsors.

Grateful acknowledgement is made to Richard Erdoes, Archie Fire Lame Deer, and Simon and Schuster for the use of excerpts from their publication Lame Deer Seeker of Visions.

Special Thanks to Paula Humfrey for the illustrations in this book.

PUBLICATIONS in which portions of *Prejudice Against Nature* did or will appear: Environmental Education Report, Journal of Instructional Psychology, The Cultivator, Maine Life, Our Classroom is Wild America, Tucson Daily Citizen, New England Outdoors, The Interpretive Naturalist, World Future Society, Education for Tomorrow's Environment, Education for the Year 2010, Across the Running Tide, Bangor Daily News, Maine Audubon Schoodic Newsletter.

Other books by Michael J. Cohen:
Our Classroom is Wild America
Across the Running Tide

Copyright © 1983, 1984, Michael J. Cohen
Second Edition

COBBLESMITH
Box 191 RFD#1
Freeport, Maine 04032

ISBN #0-89166-016-x

INTRODUCTION TO SECOND EDITION

by Marshal Case, Vice President and Director of Education, National Audubon Society

"Prejudice is a form of blindness." "Our prejudice against Nature has been covered up—justified—as growth, a conquering, progress, security, dominion, rise in the standard of living, prosperity, luxury, and increased economic gain. "... to deprive the planet of its life support is an act of prejudice against Nature. From our ecosystem we arrogantly borrow food, air, water, and life—often returning poison or nothing immediately useful."

Prejudice Against Nature is a powerful challenging look at self and planet with an expression of thoughts and ideas that *should* be common sense. Unfortunately, most of us are under such a controlled way of life with technology as a guiding light that we have lost contact with the natural world.

Because the Audubon Expedition Institute, guided by Mike and Diana Cohen in concert with a small dedicated staff, operates on the premise that the environment is the most powerful of educators, and since the outdoors is their classroom, these ideas are well grounded through continued assessment and first hand experiences.

As stated in the subtitle, this is a *guidebook*—not a text. This is why there is a powerful message throughout the writings. You have many opportunities for self evaluation, for agreement and disagreement, and for establishing new direction and commitment in looking at the planet as a living system.

"We Americans will remain conflicted and prejudiced against Nature until we begin to cope with the knowledge that the planet is alive, and relate to it feelingfully with a love and reverence for life and with the respect that is due one's mother." Applying the WL, or Whole Life factor, to daily human life is an extremely important concept within the following pages. Seldom do any of us look at or think about the affect our life styles have on others and, particularly, on the living planet. What are the *real* costs in providing a package of "perfect" tomatoes for consumers at the local grocery store? When a light switch is flicked on (or off) what is the *real* significance to the planet? And, do we consciously question the effects of our personal lives on the local and global life system? Should we—and will it make a difference to this living system of which we are all a significant part?

One important continuing challenge to all of us is to think globally and act locally. That, in fact, is the current underlying theme of the National Audubon Society. "Our ignorance is something to keep in mind as we make our decisions about how to exploit the natural world in order to maintain a standard of living." As a professional educator, I see that as one of the most provocative statements in the entire guidebook. We have all grown too far away from the basic *real* living system. Technology has taken over, people have little contact with natural systems, and in a race for personal materials and "things" we continue to ignore the impact on the source of everything in our life—the living planet.

This guidebook should be required reading for all educators everywhere—for those who have been in the business of education for many years, for those who are about to be in the influential position of working with students, and particularly for those who are studying to be educators. However, I believe this guidebook reaches far beyond the professional educators and its messages should appear on the tables of legislators—who vote the decisions, as well as on the tables of the voters—who direct legislators on how to vote. The message is clear—the living planet depends on it.

CONTENTS

Synopsis of Contents: An Overview

By Gene Boyington and Herb Cleaves

Our Classroom is Wild America was the title that Mike Cohen, Ed.D. gave to an admirable 1974 book about the National Audubon Society Expedition Institute program, a unique schoolhouse on wheels that he directs. The school still consists of small groups of inquiring graduate, college and high school students who form a close knit expedition community. They travel/camp across America studying environmental issues. They question people and places in order to develop practical awareness of relationships within and between their nation's ecosystems, people, and cultures. When I read it, I found *Our Classroom* to be a graphic compelling tale about an exciting learning process, but was disappointed in that it only lightly touched upon what was being learned about the country. What underlay the experiences that distinguished these students from their contemporaries at schools throughout America?

Now, in a new, impressive, and unusually wide-ranging work, Cohen places some sturdy intellectual planks across the perplexing gap between people and nature, lessons learned from his 15 years of adventurous study throughout American wilderness areas and sub-cultures. He combines the information derived from research with that acquired by experience, to present comprehensive memoirs and conclusions from a notable quest. In this pioneering study, Cohen reminds us that because most Americans have never discovered the source of our conflicts with the natural environment, we have been unable to rectify our continuing social and environmental problems. Instead, we focus on our immediate troubles while underlying disorders gain strength. Cohen asks, *"Why do we compulsively over-exploit and pollute to our own and the planet's denise?"* He provides crucial insights and remedies to this question long left hanging.

Our ordinary ways of talking about ourselves and other people, or of justifying our behavior expresses a certain conception of human life that is so close and so much a part of common sense, that we can hardly see it. Each of us is a mind containing beliefs, fears, hopes, motives, and desires that cause and therefore explain our actions. These perceptions are our identity. Perhaps Cohen's most valuable experience is that his profession has immersed him in the outdoors in intimate contact with nature and other peoples. Unbeknownst to himself, this setting partially weaned him of his dependency on ordinary ways to justify or explain himself. In time, an acute new awareness emerged of how and why he and others related to nature. It is fascinating that during the 16 years he was separated from the world of literature and research, new studies appeared that tend to confirm some of his observations from his backcountry adventures. Thus his experiences contribute to concurrent validity, one of the strongest tests of truth. In turn, recent literature substantiates his thesis that we're prejudiced against nature, and that we must deal with the problem accordingly, if we're to achieve a more balanced existence.

As it is drawn from nature, Cohen's awareness becomes a global consciousness that motivates him and others to act on local issues. He suggests that thinking which integrates scientific research with inborn senses and feelings is necessary, if true cooperation between people and an equilibrium with the planet is to prevail. We must be able to hear the ecosystem, if we are to harmonize with it.

Much of the author's compassionate quest is based on his personal contact with the landscape. He insists that anybody that had such exposure to natural systems would come to the same conclusions about America's relationship with the global ecosystem.

Scientifically, Cohen's dynamic social criticism blends facts and feelings when he investigates the origin of our self-destructive relationship with the ecosystem. He solves a profound mystery by illustrating that our environmental dilemmas are due to an unrecognized psychological distancing-from-nature phenomenon which occurs during our maturation from human embryo to adult.

Researchers have long known that the prenatal fetus in its relatively secure mother's womb, on a feeling level subconsciously remembers the womb environment as a supportive utopia. Cohen suggests that this lingering emotion is the source of most people's security and euphoria fantasies. In contrast to it, after birth, the periodically stressful postnatal environment of "nature" compares unfavorably. Replacement securities are constantly sought, but the process opens a Pandora's box.

Most American children are taught that technologies can provide security. For example, a house can offer womb-like fortressing from nature's fluctuations of temperature, wind and weather. Psychologically, a biased technologic security, anti-nature mentality evolves. It is affixed by the average American spending over 95% of their lives indoors, away from nature's fluctuations. Cohen finds that anti-nature childhood attitudes and their reinforcement are the underlying cause of our social and environmental problems. They run amok because they're a subconscious force that we have not yet recognized, nor properly symbolized. They taint much of our culture and contaminate our objectivity. But what is eminently important about this dynamism is that once acknowledged, our prejudice against nature can be subdued and real security gained by relating directly to people and by learning to appreciate the planet's life-giving fluctuations.

Recounting undeniable experiences, the author responds to the criticism that the planet (nature) is not alive and therefore we can't be prejudiced against it. He convincingly demonstrates that nature is basically a fluctuating, tension-building and tension-releasing process (T-R) between the entities of the universe. He reasons that human feelings are not expressions of the ego as is commonly believed, but are expressions of nature's universal T-R process.

Unholistically, our culture teaches us to avoid nature's tension. To attain "security," we attempt to stabilize nature's pulse and in the process subdivide the life systems of the planet and ourselves. We perceive the planet to be mechanical, frightening and without consciousness. But Cohen wisely observes that when we so treat the planet, it looks and acts dead, yet if we give the planet the care we would accord an animate living being, the planet responds as if it is alive.

In contrast to mainstream America's dead planet perception, Cohen documents with compelling insight why throughout history, the fluctuating, nurturing Planet Earth has been recognized by many to be a living organism—in fact, in many cultures it is recognized to be the womb of life. Cohen demonstrates that ecologically and emotionally, humankind and Mother Earth are as congruent as a heart and a person. Humanity has evolved *with* the Earth and is an organ of it. *Both are integrated life systems.* They communicate; holistically or exploitively, what happens to one usually happens to the other.

Out of his personal early experiences, Cohen perceives that he had formally been unable to include this concept in his everyday thinking and identity. This, he explains, is because our society failed to recognize or teach him that nature does not only communicate in technological, scientific terms. Nature often speaks through feelings, and since they are natural, in our culture they're avoided; they're demeaned as being subjective, if not subversive or almost taboo. In this way, we have learned to ignore nature's voice.

Prejudice is a dirty word. Convincingly, Cohen has avoided its profanity by unfrocking the forces that evoke prejudice against nature. He pinpoints their adverse impact on people and the planet. He demonstrates how, like excessive technologies, the symbols and images in our mind and language often create distance from nature's fluctuations. By demeaning our feelings, too often we relate only to and through our culturally biased symbols and unwittingly impose them upon the global life system.

To counterbalance the cultural prejudices he's inherited, Cohen has devised a Whole Life Factor that he invokes into his relationships with his social and natural environment. Through it he locates and deals with the forces that distort our thoughts and desperately need our attention. It is nothing less than a frontal challenge to the excessively technological answers offered by many scientists, economists and educators. It's a paradigm that has the markings of what could be a socio-environmental revolution.

Seeking excessive "technological euphorias" that counteract nature's fluctuations is an anti-nature tendency at birth that prevails only because it's reinforced, says the author. He describes how the material and social environment condition our inborn survival desires into destructive but culturally acceptable habits, status and a false sense of security.

Uncaring attitudes about life are side effects of our prejudice against nature, says Cohen. In comparison, he shows that they're found to a lesser extent in the hunter-gatherer cultures because their way of life "progressed" by constructing a social setting that achieved a harmonic equilibrium with the planet's fluctuations; we, on the other hand, "progress" by technologically distancing ourselves from nature's pulse, usually to our own and the planet's cost.

Cohen's Whole Life Factor is a means to let Mother Nature confront our prejudice against her. For impact, he lets her actually do so and she shines in the endeavor. In other hands, her complaints might remain dry social psychology, but here her spirited voice speaks with a sharp tongue. She claims that as exemplified by smoking health-hazardous cigarettes, our acculturation into mainstream America can all but nullify the protective self-preservation feelings that nature gave us. This causes in us a lack of full concern about health hazards, such as our cigarettes, poisonous air and water, toxic wastes, radioactivity and acid rain. Mother Nature says that we knowledgeably and uncontrollably abuse the environment and ourselves because, through our prejudicial upbringing, we have lost our natural motivations and abilities to do so.

In a fascinating dialogue, at times humorous, Cohen lets Mother Nature critique his ingrained biases and shortsightedness. Her tirade exposes prejudices in him that exist in most of us, but which we seldom recognize. She explains that each of us has two mothers and that she's one of them, that the security of the human womb was and is her and her global life-supportive harmony. She states that the underlying cause of modern war is that people don't recognize that they're at war with nature.

To reverse our anti-nature conditioning, Cohen has implemented a new accredited curriculum at the Audubon Expedition Institute. There he and his staff let direct contact with natural processes teach people to perceive their unacculturated human nature and feelings at birth to be identical with, and often as exploited by culture as is the Planet Earth. His students experientially learn to invoke a Whole Life Factor into their relationships with the planet, institutions and individuals. This academically sound approach has been shown to strengthen wholeness and self-preservation feelings in students by reducing their internal and external conflicts with nature's fluctuating processes. It is reportedly therapeutic, reduces apathy towards the political process, and retards alienation.

From his experiences, Cohen describes how the average American's surroundings limit his consciousness of nature and imbue a deep discrimination against natural systems. He says that until now, most American religions have been fingered as the cause of our environmental problems because, in the Bible, God tells people to subdue the Earth. Cohen argues that religions are cultural institutions, and as such they are subject to our culture's prejudice against nature. It is our culture's prejudice, not our religion, that is environmental culprit. Specifically, he shows how invoking the Whole Life Factor into thoughts and actions produces an environment that can implement planet-person congruency. It has already led to promising legislation that encourages the assignment of a numerical planet-congruency factor to products and services.

Necessarily, Cohen's data is gained from natural and backcountry areas instead of from the social mainstream. But the conclusions he draws are confirmed by research that Jim Swan, Ph.D. presents in the final chapter. Swan and the author's students hold a two-week research seminar and produce some valuable new definitions and goals for environmental education.

Commentary:

Cohen and his colleagues at Audubon are desperately trying to improve our life on Earth. The task is not an easy one. His book strains at the limits of mainstream logic because as he writes, a paradigm shift emerges as potent as any that has been suggested. He says, "I mistrust many answers I have received from our institutions. I equally mistrust some questions that my culture has taught me to ask. Exploitive, anti-nature questions have produced answers in me that are profitable in the context of the culture, but are terribly wrong for the survival of human life. Too often, we have earned good grades for learning how to hurt ourselves, and the situation continues to deteriorate."

The author is the book's central figure, a man of some years, shaken by the fragile beauty of vanishing life relationships. Although his book can be aggravating, it's intensely personal in its expressions of the exasperation, despair and pluck that people's interactions with nature evoke in him. His writing's peculiar, redundant sincerity is its strength, for Cohen heeds an inner voice whatever the cost. But his overall goal and means of promoting person-planet congruency is not fantasy. For global and personal health, we need to invoke into our lives a Whole Life Factor in order to erase our prejudice against nature and restore harmony to the world.

*Portions of this material are from the Bangor Daily News.

Foreword

What A Way To Learn

A description of the environment
in which *Prejudice Against Nature* was conceived.

On a fading October afternoon, an ordinary-looking yellow school
bus wound its way through the Downeast Maine countryside...ordinary
except for the fact that atop the roof were strapped 24 backpacks. In the
modified interior, perched precariously on the wooden food bins, Sue was
tuning her fiddle. She was joined by Bobbie with her concertina, and
Rachel with her mandolin. Serena reached for her pennywhistle and
arrived in time for *Sailor's Hornpipe*. Diana, in front of the bus, was
finishing a handknit sweater. Alex was charting our course on the map.
Mary and Benjamin were deep in conversation, and I was watching the
changing landscape. Without missing a beat, the musicians segued into a
reel. I dug my spoons out of the box under my seat and tapped tentatively
along. It was an ordinary-*looking* yellow school bus, but its purpose was
not. We were not on our way to school—we were there.

We were a part of the National Audubon Society's Expedition
Institute—a travelling community of students learning not from
textbooks, but through experience.

One of our first lessons had taken place on a tiny, spruce-covered
island in northern Maine. We had canoed against a pelting rain to get there,
had set up camp in the deepening darkness, and when the sky cleared had
gathered on the granite shore to study astronomy. Mike, one of the school's
guides, asked us to think of the sky as a roadmap. Starting at Arcturus, we
began to travel across the winter stars.

Suddenly, a shriek pierced the night. Someone on my right gasped,
"Look! Over there!" I turned to the sight that now held everyone's
attention. Shifting, folding layers of tenuous color danced gracefully in the
black night. Shafts of green shot upward and drifted slowly down again.
Curtains of white yawned and stretched. Translucent blue darts shot out in
continuous exclamations. The whole scene pulsed kinetically—a ballet of
light and color. Each second brought a new wonder, a new arrangement of
pearl and sapphire and tourmaline green against the ebony backdrop of the
sky. This was the Aurora Borealis as it had rarely been seen in this part of

the country. We settled back, relishing the display. Our astronomy seminar had taken a twist of its own.

Audubon's Expedition Institute (AEI), is an accredited one-room schoolhouse on wheels. In September of 1980, I joined it for an eight-month trek across America. Our class was composed of two guides, several graduate students, many undergraduates, and two high school students. We ranged in age from 15-50, but all were there for the same reasons—to form a self-governing community and to learn. We didn't have textbooks, but carried a small library in the back of the bus. We didn't have scheduled classes or exams of any kind. And each night we slept on the ground—sometimes in tents, often under brilliant fall and winter stars.

The subjects we studied were traditional, if the methods weren't. Our teachers were the people we met, the places we travelled to, books in libraries all across the country, and the vast knowledge of our guides, Mike and Diana Cohen. AEI was far from a collection of unrelated academic subjects. It was an opportunity to discover that all subjects are intertwined, a chance to find new values in living and learning, and to explore the immense land and diverse people that make up this incredible place we call America.

In the 1800's, there were many idealistic communities in existence in the U.S. All of them felt they had found a better way of life. Few of them remain today. We visited one—the last working Shaker community, in Sabbathday Lake, Maine.

We were invited to join the Sunday Meeting, but were informed that most of the community would be leaving right after the service, and only Brother Arnold would be available to talk to us. As we entered the main house, through separate doors for each sex, I noted the beautiful wood floors, the clean, austere rooms, the absence of decoration. We took our seats in the meeting room—men on one side and women on the other. As the service started, I thought it was a shame that most of the community already had left. I learned later that they hadn't. The Sabbathday Lake Shaker community has only nine members. They all had been present.

It was Brother Ted's turn to deliver the short sermon, after reading a passage from the Bible. The topic was "forgiveness." Each person was then free to respond, adding his or her own thoughts or applying it to his or her own life. There was no dancing, none of the famous "shaking" from which they'd gotten their name, but occasionally someone would start a song. *'Tis a gift to be simple, 'tis a gift to be free.* In a quiet moment, I noticed that the first snow of the season was falling softly outside.

Later, Brother Arnold was happy to answer our questions. The first on everyone's lips was, "How could a celibate society have lasted over 200 years?" Brother Arnold explained that the Shakers rely on converts, adding that he had been a member for only three years. I asked why he, a goodlooking man in his early 20's, would choose to isolate himself from society and live a celibate life. "Because I believe in it," he answered simply. "This is the way to live."

Before joining AEI, I had little understanding of what an "expeditional" school would be like. This is what I found:

Expedition meant a group of diverse, motivated people working toward a common goal, in this case knowledge and perpetuation of the expedition. It meant commitment. We didn't come and go as we pleased, we considered the effects of our actions on the functioning of the group as a unit—we made sure that everyone had firsts at dinner before we took seconds. It meant finding a way of making decisions that didn't leave out anyone's feelings. For us, that way was consensus. By discussion and compromise we made sure that every decision was one that every person agreed with.

Expedition meant communication. We were at school 24 hours a day, 7 days a week. The walls erected for protection elsewhere were barriers to smooth relations here. We learned to express our opinions, confront others instead of ignoring them, talk if we were troubled. The result? Not a single person was, or felt, unneeded. Each of us knew our opinions were valued, our well-being essential. We found that others liked us, even at our unpretentious worst (best?). We felt closer, formed relationships stronger than many of us had ever known before.

On a hike to the majestic Rainbow Bridge in southern Utah, we devoted an afternoon to swimming in one of the most beautiful places in the Southwest. It always surprises me to find a rippling stream in the desert, but on this hike we had criss-crossed one for days. Here the river narrowed and bounced down steps of carved rock. And it formed pools—each one larger than the one behind and above it, from depressions not big enough for both feet to bathtub-sized reservoirs to pools we could actually swim in. And the water changed magically from crystal emerald green to royal, rich aquamarine blue, always strikingly framed by the hard, grey rock it sculpted.

We played unself-consciously. I loved being so childlike in this setting, making body dams to hold back the water, then sending it spilling down on someone below. We played with delightful abandon, diving into the long, deep pools of icy water, daring each other and not even caring if the dares were accepted. It was fun. And when we sat on the sloping, slippery rocks to rest and share someone's carefully guarded M&M's, the sun dried our shorts and T-shirts so that we had to peel them away from our skin.

It seemed, that afternoon, that we had finally come to an acceptance of each other, an appreciation of each other's inherent worth, and a special kind of love and caring for each other and the planet that naturally followed. It was a good, satisfying feeling. Sometimes I learned more facts than I wanted to know. We met an old horse-and-buggy Mennonite family in Pennsylvania who needed help on their farm. We swapped ten days worth of working for them in order to share ten days' worth of their 18th century farm life. We gathered eggs, built a wagon and cleaned sunflower seeds, which gave the farmer time to meet with us and answer our questions.

One day when his daughter wasn't watching, the farmer took us into a special room in the farmhouse, the parlor. He explained that it was the courting room and told us that, "by all of the Plain People it's the same," and that "as is common by us" his daughter only had her boyfriend visit her in the parlor on Sunday night promptly at 7:00 p.m. When her boyfriend left at midnight, he could sleep in his buggy on his seven mile journey homeward, for the horse knew it's way across the familiar backroads to the young man's farm. "Yes, that happened also with me when I was courting," said the farmer. Later some of us rode in his buggy after learning how to hitch up the horse to it.

Although the Plain People's low-technology-lifestyle is old fashioned, it was new to me. It almost hurt when the farmer mentioned that the harmonious combination of their traditional outlooks and farm techniques had kept the land as healthy as it was 250 years ago. For me that discovery was also the discovery of a deep feeling of concern. It led me to question my relationship with nature, and the effects of how I'd been brought up.

For me, finding AEI was an exciting alternative to traditional classroom settings. I learned to take control of my own education—anywhere—rather than let someone spoon-feed me. I became aware of the effects of my actions on others and on the natural world. I began to make decisions about my life based on rational thought, and to act in ways that made sense to me.

And what about grades in this unconventional school? Who graded the Expedition members? By now you've probably guessed. We graded ourselves. It involved long hours of introspection, and a self-evaluation presented to the group for yet another consensus decision. But interestingly enough, it was the first time I felt my grades had really been earned. It wasn't cramming for tests, it was a daily accumulation of processes and ideas.

Expedition education is not for everyone, but for most it can be a valuable and treasured supplement to classroom learning—where lessons and facts are important because they are lived.

Foreword

It was April when we sat atop the Hopi mesas in northern Arizona waiting for the spring dances to begin. And it was cold. We had split into small groups in order not to dominate any one kiva. My group huddled outside one of the ancient ceremonial rooms sharing a down jacket among three of us, hoping that the midnight dances would start early and usher in spring successfully. Around us, anxious mothers clutched the hands of nervous, black-haired children already into their second wind. Youngsters flailed each other with paper bags—bags that would be filled with goodies when the dancing was done. Under the crystal stars, we all awaited the arrival of the Kachinas—the spirits that come down from their home in the nearby San Francisco Peaks to spend spring and summer on the three Hopi mesas.

The door to the round mud and grass kiva opened, and we were beckoned inside. There were pictures along the walls, chairs where we sat at one end, and a woodstove burning sweet, fragrant juniper. In the center of the kiva, a ladder reached up through a hole in the ceiling, and now we could hear chanting and stomping on the roof; the Kachinas had arrived. Down the ladder they came, one by one, and circled the dance floor. Their elaborate costumes were blazing with color from feathers, paint, fox pelts, and blue Hopi corn. Bells tied around their shins jingled at every stomp. These dancers were the spirits of deer—deer Kachinas. We could see four different sets of Kachinas, and even some playful mudheads before the night was over. The children, agitated and fascinated, did not know it was their fathers and brothers dancing. We all sat enthralled by dances that have endured for centuries.

Their dancing done, the deer Kachinas reached into heavy sacks tied to their sides, and pulled out handfuls of candy, blue cornmeal piki, popcorn, loaves of bread, and fruit, and hurled them at the children who squealed and grabbed and stuffed their bags in a ritual that would be repeated three more times by the other Kachinas. Already we could hear the next group of dancers on the roof. The deer Kachinas crowded around the ladder, then, one by one, climbed up and disappeared through the hole in the ceiling.

It was 2 a.m. by the time the last dancers left. Our prayers for spring rain to water the crops were over for this night. We left the cozy warmth of the kiva and drifted out into the friendly black night. The last sounds to leave my ears were the cries of the children scooping up handfuls of candy from their bags, and tired mothers shooing them along home. In another week, I, too, would be going home.

And on an expedition that was studying people's relationships with nature, a week was a very short period of time.

Chris Johnson
January, 1983

Prejudice Against Nature
A Guidebook for the Liberation of Self and Planet

Prologue

BUT ONE BREATH HAVE ALL

6:15 AM. The universe pulls the bay waters up to the seaweed line that marks yesterday's high tide. I, seeking the sources of our prejudice against Nature, push the canoe north'ard and let it gently glide with the new outgoing tide. This is the last such ride. Why did they build a tidal hydroelectric dam? Where will the water be at 8:30, I wonder? Why will the sea never come back?

I am carried like a corpuscle in a giant artery from the heart of the Earth. The futile salty tears on my cheeks are a fitting eulogy for the saline plasma of seawater that now retreats ten feet per minute from the shore. Who recognizes the bay waters as lifeblood: a reverent mixture of food, air, and minerals, stirred to perfection by wind, waves, and the timeless flushing of the one-mile tide four times each day? Who cares if the seaweed deposited at the highest tide this month would have decomposed into a unique nourishing fertilizer to be carried to the waiting offshore waters at the highest tide next month?

I am engulfed by fish, shellfish, micro-organisms, marsh grass, and seaweeds. Representatives of every pulsing, wriggling, flowering life form have gathered here to suckle the lifestream of the bay. They are sustained by it, as am I. It is no wonder that so many of us choose this place to live, for here we don't buy energy or food. Nature freely shares it with us at this and every other moment. All that is asked is that we return it when we need it no longer.

6:57 AM. The sun breaks through the easterly fog over the shoreline towns and cities. I only wish I could, too. Why do we deny and discourage life? Why did they build the dam? What is so hateful about this place that we are driven to kill it? I am looking into cold clear water. I dip my hand into a pleasant forest of rockweed, snails, and periwinkles—friends and food. The wind whispers; a gull floats by; a seal follows my canoe with curiosity—submerges, and appears again—perhaps to confirm that it is seeing correctly, for the boat does not have a motor. A loon calls, as small waves and dancing sunlight glance from my paddle. What is so boring about here and now? How is it harmful? I float. I am being drawn out to sea

17

by the powers that created me, while others are drawn to destroying these powers, to enslaving them with a tidal dam to run their television sets, instead of letting them be to run the planet.

7:35 AM. The far-flung ring of grassland, mudflats, and seaweed-painted boulders grows larger as the receding sea gives its intertidal sustenance to fish, mammals, and fishermen offshore. I watch the demise of multibillions of shoreline life forms. They are drowning by being submerged in air, just as I would drown, if I were submerged in water. They will wait expectantly for their bloodstream to return as it has for millenia past. They don't share my secret; they don't know that the sea has made its last visit. That lesson will be learned the hard way; as the slow, deliberate stress of suffocation and thirst begin to be felt; as desert dryness and summer heat leisurely broil the flats and their inhabitants.

Have I gone mad? Am I the only one who recognizes that human consciousness is shared, at some level, by this community from which we gained all of our life processes? Am I deceived in believing that the turmoiled intertidal community is conscious, that it knows that seawater is but inches away and that it will not return on its appointed rounds? How convenient for us to conceive mud, water, and stones to be dead; to decide that other life has no consciousness, pain, or equality. What an incredible alibi we have created to soothe our guilt of killing for a profit, of brain-washing our neighbors into believing that they are *nothing* unless they brush their teeth electrically. Isn't it the economics of a madman to trade away a life system for an electric can opener? I murmur sadly to the bay, "We hate you."

8:45 AM. The tide has retreated to the highest point it will ever again reach in this finger of Cobscook Bay. Hundreds of acres lie exposed to an environment as hostile as that of the moon. The inconceivable trade-off begins as the minutes tick by. Where are the friends of the Earth; where are those Americans who say they care? Are they watching television and dreaming of further riches to be gleaned from the unnecessary tidal power, while a part of the bay dies along with a part of themselves?

As if disgusted by the nightmare it leaves behind, the tide retreats below its new highwater mark. But as the Maine State Legislature has proven, this is no dream. A Tidal Power Authority has been authorized. One wonders when we will outgrow our childhood. One wonders if our immorality to each other begins at the interface with Nature's far-reaching shoreline. One wonders if the tide somehow says goodbye to its ancient home.

Chapter One
ENGAGEMENT WITH CULTURE

*I, seeking the sources of our prejudice against Nature, push the canoe
north'ard and let it gently glide with the new outgoing tide.*

I, my wife, and the Expedition students were sitting wedged in a tiny
living room in Arizona, listening intently.

"They wanted us to learn white ways. They brought in the U.S. Army
to make us go to school." The old Hopi woman's voice was soft as she
stared at the floor and remembered her childhood. "But still my parents
and other Hopis resisted. The soldiers, every morning they came with guns,
went from house to house, searching for children—so every morning, very
early, one of my people would gather us children away to hide..."

Society in the United States traditionally has viewed the Native
American as an anomaly: a race of savages, not quite human; a feral and
cunning people living in primitive conditions, close to the Earth. We have,
in the same tradition, done our best to change that. We have taught them to
listen to our white noise.

"One by one we were found and caught," continued the old woman.
"Children and parents would be crying; the soldier would have one of your
arms, your mother, the other; and you were pulled away, kicking, from
everything you know, to go to the Indian school. You did not come home
again for twelve years. I was one of those children."

They were Indian children, angels of Satan, the uncivilized children of
Nature, and they were forcibly educated to our ways.

"We did not have our lessons in Hopi; they were in English. Hopi was
forbidden. We did not understand English, but we were punished if we
were heard speaking Hopi among ourselves. The principal had a pistol full
of blank cartridges. He used to fire it over the heads of the bigger boys to
frighten them into obeying the rules. The food was strange to us; but if we
did not eat what we were served, we went hungry. We were not allowed to
go home in the summer for fear that perhaps we would not come back—so
we were put to work at summer jobs. Anyway, after a few years we would
not fit in at home. We were changed."

The living room was taut with amazed silence, broken only by the
nervous laughter with which the Hopi woman punctuated every statement.
She showed no resentment of the treatment she and other Hopi children
had received. The only sign of her discomfort was this recurring, apolo-
getic titter, as she looked up from the floor and read the horror in our eyes
that reflected some long-ignored part of herself.

We remembered the Hopi schoolteacher who had told us, "American education is the weapon that has finally invaded and conquered the formerly invincible Hopi, a peaceful people who once successfully repelled both Coronado and Christianity." Now we understood.

The old woman gestured around the room, at the three television sets, the stereo, the bookcases full of books, the couch, the electric drier, the coffee table, the wall-to-wall carpet.

"Now I am an American," she said proudly. But there was a crease between her brows—and she laughed.

§

As co-founder and director of the National Audubon Society Expedition Institute these past fifteen years, most of my nights have been spent sleeping under the stars, winter and summer. Surroundings of unfamiliar habitats and sub-cultures have been my home. My companions have been my wife, my dog, and my extended family: a small academically accredited travelling community of graduate, college, and high school students, banded together as hunters and gatherers of knowledge. As a group, we have investigated the relationship between people and Nature. Our laboratory has been real-life encounters in hundreds of American cultures and ecosystems from Maine to California. We have sought some of the secrets of American life.

The Institute operates on the premise that the environment is the most powerful of educators. For good or bad, the environment teaches us to be what we are. Our language, our outlooks, and our way of life: we are the sum of our genetics and our interactions with the different surroundings of our pasts.

Every environment moves into the perceptions of its occupants. People in any society ride the carousel of their culture. We go around the limiting circles that exist within our frame of reference, often unable to establish new directions.

To independently change life directions requires access to new perspectives. Dismounting from the carousel donkey merely to mount, instead, the nearby elephant, changes very little. Completely disengaging from the carousel and observing it without being subject to it gives us a much clearer viewpoint and relieves dizziness.

The Institute is an attempt at education through disengagement.

During the twenty years that I myself attended kindergarten through graduate school, I was taught how to make a living *but not that the planet might be alive.* I learned the importance of progress and technology, *but I was seldom taught the importance of Nature.* I learned about living with people, *but not how to live with the planet.* I learned the value of money, *but not the value of my feelings.* I was taught how to adapt to society, *but the civilization of the wilderness was never in the curriculum.* It is the years since I left school that have integrated these complimentary bodies of knowledge, for during these decades Nature has been my teacher.

On the yearly expeditions, and in these pages, concepts are discussed which have been gleaned from a constant encountering of the variable mosaic which we call the American environment. We have listened to the Amish in Pennsylvania, the Hopi in Arizona, and the rainy wind soughing through the trees of the Smoky Mountains of Tennessee. We have heard politicians, musicians, historians, and the gusty spume of humpbacks breaking the surface of the ocean. Farmers, prairies, lakes, and lumberjacks—we have learned from them all.

As we've travelled across the country, our group has communicated with many Native Americans. In our conversations with Passamaquoddy, Papago, Seminole, Zuni, Wampanoag, Yaqui, Hopi, and Navajo, we have been impressed by the integrity of their prehistory cultural experience. For thousands of years their ancestors lived in almost seamless equilibrium with Nature, revering her as the Great Spirit. They had no word for *wilderness.* If their languages had contained such a word, it would have been *home* or *life* or *us.* Although today most Native American cultures have been overwhelmed by technologies and media-sown perceptions of reality, we have heard, in quiet moments, Nature speaking through the voices of these people who once lived close to her.

John (Fire) Lame Deer, a Sioux medicine man interviewed in the early seventies by Richard Erdoes, communicates this ancient Native American perspective in this book. I have purposely chosen to include the thoughts and self-preservation feelings of this one man rather than the Indian thoughts and feelings of a scattering of individuals we have met. Within a single person, the conditioning of a culture is imprinted on the framework of a living being, making him an acculturated whole. In our thirst for academic interpretation and scientific objectivity, we tend to forget the individual and his/her innate desire to live. An individual is the die cast by his/her cultural mold. The outlooks of a specific individual can vividly reveal the parameters and values of that mold by *conveying its effects upon him/her.*

As with any other sect or subculture, the outlooks of the American Indian population are wide and varied. Some Indians believe in returning to wilderness living, while others pursue modern business and political careers. Lame Deer is particularly interesting because his ideas seem to represent the outlook of an historic hunter-gathering people who lived close to Nature. Their descendants are now subjects of our dominant culture.

Although Lame Deer had little formal education, he was a learned individual. His cultural perceptions of us help us see ourselves. As our travelling community has applied his thoughts to our own lives, we have discovered new meanings and new directions. I'll let him introduce himself:

Lame Deer:

I was born a full-blood Indian in a twelve-by-twelve log cabin between Pine Ridge and Rosebud. **Maka tanhan wicasa wan**—I am a man of the earth, as we say. Our people don't call themselves Sioux or Dakota. That's white man talk. We call ourselves Ikce Wicasa— the natural humans, the free, wild, common people. I am pleased to be called that.

As with most Indian children, much of my upbringing was done by my grandparents—Good Fox and his wife, Pte-Sa-Ota-Win, Plenty White Buffalo. Among our people the relationship to one's grandparents is as strong as to one's own father and mother. We lived in that little hut way out on the prairie, in the back country, and for the first few years of my life, I had no contact with the outside world.

When I was about five years old, my grandma took me to visit some neighbors. As always, my little black pup came along. We were walking on the dirt road when I saw a rider come up. He looked so strange to me that I hid myself behind Grandma and my pup hid behind me.

He had eyes like a dead owl, of a washed-out blue-green hue. He was chewing on something that looked like a smoking Baby Ruth candy bar. Later I found out that this was a cigar. This man sure went in for double enjoyment, because he was also chomping on a wad of chewing tobacco, and now and then he took the smoking candy bar from his mouth to spit out a long stream of brown juice. I wondered why he kept eating something which tasted so bad that he couldn't keep it down. This funny human being wore leather pants and had two strange-looking hammers tied to his hips. I later found out these were .45 Colts. He took some square green frog hides from his pocket and wanted to trade for Grandma's moccasins. I guess those were dollar bills. But Grandma refused to swap, because she

had four big gold coins in her moccasins. That man
must have smelled them. This was the first white man
I met.

As our expedition school journeys across America, we gain full
graduate, college, and high school credit for what we learn. I observe my
peers spending fortunes chasing the values that daily I live and teach, while
my modest salary accumulates, because I have little need to use it. Best of
all, immersion in those vastly differing environments and cultures has
given me a new understanding of what it means to be a whole person.

With dismay, I have discovered that internalized within me are
aspects of the American environment that I abhor. Prejudice is part of
America. As I am an American, it is, against my will, a part of me. As I sit
here thinking that prejudice is not very pretty, an old song comes to mind:

Sometimes trees bear a strange fruit
Blood on the leaves
And blood on the root
Black bodies swinging in the southern breeze
Strange fruit hanging from the poplar trees

At times I have been on the receiving end of prejudice, and as do
redskins, chicks, spics, niggers, wops, and kikes, I hate it. It is stupid,
powerful, and ugly. It hurts. It makes me angry.

I try to thwart bigotry when I find it in myself, but I find that I am able
to make headway only when I have some kind of outside support. Maybe
this is because prejudice is learned from an environment, and must be
offset by a different, more reasonable environment. Such was the case in
1961 when I started a business of my own in New York City and occasionally did not have the money to pay the monthly $115.00 rent. Mr. Kessler,
my millionaire landlord, was a pill, usually abrupt, and especially nasty
and aggravating when rent payments were late. Although I was often hurt
by his arbitrary anger and insinuations when I was broke, I could more
powerfully contain his arguments when I had the rent money in my pocket.
The money was a support system, a strength that was otherwise missing
because as a rent payment delinquent I was my own enemy. I was part of
the problem.

This book is about prejudice, you, and me. It may, at times, prove to
be uncomfortable reading because the discovery of prejudicial relationships is as emotionally confronting as was integration during civil rights, or
Mr. Kessler's hang-ups.

My wife and I have been able to deal with the prejudice this book
describes only because during the past years we have learned from Nature

to be less prejudicial toward Nature. We no longer fear or dislike Nature as we did two decades ago; we no longer enjoy maintaining an unnecessary distance from the fluctuating energies of the natural world. It is common for Diana and me to sleep outside in a tent. Our small home does not have electricity, indoor plumbing, central heating, or a refrigerator, even though we were brought up accustomed, if not addicted, to these "conveniences." Although we don't believe that our rural lifestyle is for everybody, we recognize that our lives and fun can serve as empirical evidence that many of today's "necessities of life" are not necessary for wholesome survival and happiness. We are not naive. Our lifestyle and expedition travel, in one sense, give us a scientific control with which to objectively critique America's survival relationship with the Planet Earth. We have found that without having the "rent in our pocket" of our current lifestyle, any anti-Nature acts to which we have been exposed over the past decades might otherwise be unquestionably accepted by us as a necessity in coping with life. Instead, our experience and observations raise some serious questions about why environmentally and socially our society is so self-destructive. They lead us to question the effects of our personal lives on the local and global life system. For some open-minded people, there is a validity to such an inquiry and the preventive measures it has produced, while to other people it feels like convincing the Ku Klux Klan that they should encourage Blacks to become members. We find that too many Americans are prejudiced against Nature, but that they don't recognize it. That's a danger of prejudice—it's a limited state of consciousness.

NASA space photographs confirm what Nature has taught me to read from my senses. The blue color that I see in the midday sky and the sea is matched by the blue of the planet in the photographs. The roundness of the Earth's shadow on the moon—which hints that the Earth is a sphere—is confirmed by the NASA photographs of the globe. The prints also corroborate the physical wholeness of the planet: its isolation, its motion on its axis, its whirling atmosphere, and its incontrovertible beauty.

I have come to realize that at birth, I, like everybody else, was a whole, dependent, ball of life that was colored *Planet Blue*—the color of our life in the solar system. Perhaps circumstantially, in the foetal position in my mother's womb, I was circularly shaped like a globe.

After I was born, I was, again like everybody else, painted different colors by my social environment, my culture. I was thoroughly washed with the colors of being American, Judeo-Christian, democratic, white, English-speaking, middle class, materialistic, educated, socio-economically adept, and housebroken. Each of those cultural colors captured and covered up a portion of my natural wholeness, my blueness, until I could no longer recognize my original self. Every other person on the planet experienced a similar coating process, except that each received different colors, and some did not collect as much paint as others. We tend to forget that we view the world through the filter of the colors we have been painted. Seldom do we consider the effects upon ourselves, and on the planet, of the paint jobs we received. We ignore in our immediate preoccupations the possibility that our cultural paint may be giving us lead poisoning, clogging our pores, blocking out health-giving rays from the sun, and reducing wholesome exposure to the atmosphere. It may also be creating confusion, self-doubt, and a lack of sure identity, as uneasily we question which of our colors is really *us*. The layering of cultural paint is so thorough and skillful that often we are unable to see that underneath it all we are still *Planet Blue*—still the color of the Earth and of life.

As cultural compositions, we are apt to overlook the importance of our hidden blueness and its vital congruence with the planet. To the contrary, most of our lives are dedicated to discarding our blueness and devaluing things that are blue. We expend our energies *getting and spending,* gaining and intensifying the cultural pigmentation leached from advanced technologies, economics, academics, and status. We have learned to ignore our self-preservation feelings from Nature.

A part of my own blueness is that I was born left-handed. Besieged by every possible pressure—parental and teacher disapproval, poor penmanship grades, a reduced allowance, the ridicule of peers, and lower-than-deserved grades in other subjects due to the illegibility of my handwriting—I finally managed an awkward covering over of my lefty blueness. But I never wrote well. Writing was always a subconscious conflict between blue and white, until the age of eleven when I was again allowed to write with my left hand. Although my handwriting did then slightly improve, it was too late. I've never learned to write legibly. I have never forgotten, however, what I did learn: to follow my feelings and respect the gift of my left-handed blueness.

During the years following my formal education. I have lived in unusually close contact with many other human beings and environments. Their colors have rubbed off on me; they, in turn, have rubbed off some of my colors. The effect has been to dissolve some of the color spectrum into which I was born. I am increasingly able to experience my blueness and shape my life accordingly. I spend my time trying to answer my own questions rather than those asked by and for institutions. I actualize curiosity and freedom, and approach equilibrium with the environment. As layers of my childhood conditioning and education have been identified and realigned, I have uncovered facts of life that most Americans have never known.

Caution should be observed when reading the chapters that follow. Like the old Hopi woman who was Americanized as a child, the reader may feel personally confronted or attacked as the adverse effects of the familiar layers of his or her conditioning come into consciousness. That's exactly how I felt as my blue planet feelings began to make sense. Nature appeared to be hostile, arrogant or self-righteous, just as Martin Luther King was perceived by many whites, or as a liberated woman is viewed by many men.

As I began to make sense of my direct contact with outdoor America and its innate peacefulness, I began to enjoy the feelings and economic benefits of camping out all year round. This in turn gave me the strength to cope with the seemingly foreign requests that Nature began to make. Nature had me beguiled, and it felt terrific.

This book was conceived and written through blue eyes. Its theories and observations are concepts that Nature has taught me to understand and live with. Therein lies its significance and value; for deep within us, each of us is blue.

Author's Note: Herein I have attempted not only to narrate events, but also thoughts and feelings associated with them. That's how the mind seems to function when surrounded by trees and mountains instead of by the logic of our civilization.

In the chapters it may be at times difficult to isolate my conclusions because they're holistically intertwined with the environments and personalities that produced them.

For this reason the table of contents contains an abstract of the book and I've highlighted these points in chapter summaries.

Summary and Conclusions

Because it often ignores Nature's global community, American education,—the acculturating process at home, school, work or play—is often destructive; however, learning directly from the natural environment has proven to offset this phenomenon. Comparing our society to a hunting and gathering culture reveals some important differences between the two, as well as the relative health of their respective environments. American prejudices are an emotionally charged part of American acculturation; their discovery in oneself is disquieting. At birth people and the planet are congruent. That we are prejudicially taught to demean the inborn natural aspects of ourselves creates hurtful planet-person conflicts. These can be counteracted by re-discovering nature's deep values in the environment and ourselves.

Chapter Two

WHAT GOD HATH WROUGHT

Who recognizes the bay waters as lifeblood: a reverent mixture of food, air, and minerals, stirred to perfection by wind, waves, and the timeless flushing of the one-mile tide?

As I had in past years, I met the thin line between love and hate in a telephone booth one afternoon. This particular telephone, which stands in the middle of a sand dune on the Navajo reservation, seems tuned to the culture that surrounds it.

Traditionally, the Navajo and Hopi have had no written language. Their myths and history have been subject to small changes year by year, as circumstances affected their oral communication.

The procedures for successfully making an oral communication from this telephone are subject to changes by the hour. But I had to call the headquarters of the Expedition Institute in New York City for my messages, so I invested my dime and took my chances.

Luck was with me—I got a dial tone. I dialed operator and, through the ear-shattering static, was informed that she could not hear me. There was an exchange of unintelligibles and finally she hung up. I inserted another dime and obtained the same results.

My technological prowess took over.

Years of brutal experience with this telephone prompted me to shift into trial-and-error exploration patterns. I dialed the number of the booth, got a busy signal, hung up, deposited another dime, and dialed operator again—quickly, so that the telephone had no time to recall its taste for static. Bingo. I got the operator loud and clear. She placed the number for me, and I deposited money as she requested. The phone slyly jingled back my quarters and dimes as the operator notified me that I had not deposited them. I explained that I had, but that the phone was evidently out of order. She insinuated I was lying. We sidestepped the issue by placing the call collect.

Anne answered at the New York Audubon office, accepted the call, and relayed a message to call Ms. Susan Gould about an article I had submitted. (Susan Gould is a literary agent, recommended to me because she is known to be successful in placing articles about the environment. The article which I had sent to her for consideration was Part I of a two-piece essay about a canoe ride I had taken on a tidal bay.)

I thanked Anne, hung up, and began the game of telephone roulette again. This time I could not get the dime to stay in the phone, but the operator came on anyhow, and we resolved the issue of payment with the use of a credit card.

A busy signal rewarded these efforts.

Two days later I called from a more responsible telephone in Flagstaff, just outside the reservation boundary. Although the Hopi and Navajo have religious claims in the Flagstaff area, which have restricted growth of a local ski development on their sacred mountains, the telephone there seemed immune to their more natural communication auras. After a series of purposeful clicks, I was talking to Susan Gould's secretary, who asked that I make an appointment to see Ms. Gould when I returned to New York in April.

At the end of my April interview, I was sorry I had taken the trouble. I walked into her office carrying Part II of the tidal bay article in a manila envelope. But no, she could not place Part I, *But One Breath Have All.* No, she did not believe it was publishable. She was a series of one *No* after another. Magazine and journal editors in the fields of science, religion, history, psychology, education, conservation, environment, and yes, even science fiction had convinced her that additional inquiry would be fruitless.

She sifted through the rejection slips littering her desk. "It does not fit our publishing schedule or requirements." "It is not in our field." "It is not scientific." "This is a story, not a study." "It is impractical for our use." "You can't change culture." "Technology is our lifeline." "It is emotional, not rational."

Susan Gould paused and looked up at me quizzically. Evidently she expected me to call a halt to this rejection. When I did not, she shrugged and went on.

"It is full of cliches." "Our readers won't go for something that criticizes their lifestyle." "Try to be positive, or pious and philosophical, like Thoreau or Wendell Berry." "By definition, people can relate prejudicially only to people. You can't be prejudiced against inanimate things like Nature." "Our magazine generally runs more upbeat articles." "The Earth is not alive like people are."

As her right hand nervously twisted a strand of her hair, Ms. Gould asked, "In light of these responses, why is it so important to you that the article be published?"

What flashed through my mind was, "think globally, act locally. How does one communicate that?"

There was dead silence, for how long I can't say. I sank back in my chair and collected my thoughts. "It feels right," I responded, "that's the main reason. I have seen so many different people with different backgrounds who are trying to find a strong meaning to their lives, to heal themselves, to find a peace or harmony globally or on an immediate personal level. Something is missing for most of us. I'm positive that I know what it is. We're all blind to life's wholeness. Like the blind men and the elephant, each of us is touching a different aspect of life—and somehow we're missing its commonality—so we argue and fight over our religion, politics, nationalities, economics, possessions, and racial beliefs. We are brought up blind to the whole; and we are enervating and destroying ourselves and each other, while trying to promote our own brand of blindness and half-truths. I know. I know because I can't deny my experiences, because year after year I live with people from all walks and outlooks of life, and as we discover that to which we have been blind, we are able to resolve the differences that we bring with us. I have actualized a communication of self-preservation that bridges the gaps between people and the planet. I've written about the process in other books, but only recently have I been able to identify the entity to which we are blind. By identifying it, I've found that on the expedition we've been able to discover some of the harmony and cooperation that we've been looking for and are able to share it with others. Isn't that important? Shouldn't people be able to learn about it from an article?"

"How does your article do that?" she asked, her voice reflecting disbelief.

"It trys to get at the heart of the matter, which is our relationship with Nature. The relationship is strained, and therefore our lives are not harmonious."

"What makes you believe that's accurate?"

"Well, on the expedition we've lived with and studied communities in America where people do or did get along well with each other. One thing that harmonious communities seem to have in common is a long term healthy equilibrium relationship with Nature—and when that relationship becomes strained, so do the community's interpersonal relationships. Most harmonious communities seem to be able to hear Nautre's voice and to heed it. We've found that be emulating some of their simplicity and life processes, our expedition community also has achieved harmony and cooperation, and our participants have gained a greater peace of mind. Most harmonious communities do not have high crime rates, drug use, or broken families, and they remain harmonious, until they replace their good relationship with Nature with extraneous technologic dependencies that are foisted on them by the mass media and big business. Once they begin to depend on mass culture, they develop the problems that are typical of our national way of life. What we've found is that the essence of harmony is attained by relating harmoniously with Nature, that the planet should be the role model for sustaining our lives, but instead that we destroy it and are prejudiced against Nature.

"Is that what you think your article says?" mused Ms. Gould.

"As best I can use the written word," I replied. "There's a second section to it..."

"Perhaps you're biting off more than you, or anybody, can chew. Maybe you're not a prophet, or perhaps words can't replace your real-life experiences," said Ms. Gould.

"Do you think you understand it?"

"Not really. I think that our society is already highly cooperative. We even have a standardization of parts and measurements; businesses in California use products from Virginia, people in Maine eat oranges grown in Florida. And you really lose me, when you start talking about listening to the *voice* of nature."

"I'm talking about harmonious cooperation, not just typically cooperating in order to be more powerful exploiters. Here, let me diagram it for you by drawing two donkeys that are tied together, and two piles of hay that they want to eat.

"Isn't that what you mean by cooperation?" I asked Ms. Gould.

"Why, it's simplistic, but...yes."

"But it's not harmony; it's not complete," I stated. "What do you think panel number 5 would look like when there is no hay left, because it's all been eaten? That's the point in time where we are today—there's a lack of resources. Without any hay, the donkeys can cooperate, but to no avail. In order for harmonious cooperation to take place, you have to go back to panel 2 and include the voice of the hay, the voice of Nature, in the discussion. The self-preservation of the hay has to be part of the solution. The message from the hay must be heard, if harmony is to exist—like this:

5.

The voice of Nature comes in from the system involved, the entire organism which includes the hay, the donkeys, and all of the life support systems. A donkey tuned into the reality of the entire organism in a harmonious way will know how much to eat and how to cooperate with everything else. Harmonious cooperation with Nature requires the use of human feelings and intuitions, which includes learning the ability to sense the totality of the entire system, to think globally.

We talked for a full hour. Where could I place Part I of my article? By the time I gathered my envelope and the confetti of rejection slips together to leave her office, I had an inkling of her professional opinion on that. At the bottom of a bird cage. Frankly, I had not expected much more.

I appreciated that she had done a very thorough job; but, as I told her, I felt like a shirt salesman. She raised an eyebrow.

"There was once a gentleman," I began, "who had constant headaches, dizziness, and spots before his eyes. He changed his job, saw every doctor and specialist in town—but nothing. No improvement. So, okay, one day this gent went to buy a shirt, and told the salesman that he took a size 15½ collar. The salesman looked at the guy's bull neck and got out his tape measure. 'Sir, you take a size 17,' said the salesman, very politely. 'Nonsense!' roared the customer. 'I always wear a 15½!' 'Ridiculous!' the salesman bellowed back. 'If you did, you'd have headaches, dizziness, and spots before your eyes!' " I grinned.

"Funny," said Ms. Gould, without a smile. "But as an allegory, it's just a bit arrogant, don't you think? You seem to be saying that your idea that the world is prejudiced against Nature is right, even though the world disagrees with you. The publishing world, anyway."

"The editors are correct—that is, predictable—in their responses to my article," I explained. "I don't question those responses. I question the source of their evidence, the process of their reasoning. They operate within the confines of four walls; I think it affects them. My ideas come directly from fifteen years of cross-cultural relationships with Nature. Nature tells me a different story."

"Perhaps I should submit your article to *The National Enquirer*, if Nature's talking to you," said Ms. Gould. "Should I try that?" she asked sympathetically.

I sighed. Forget it, I told myself.

I smiled, shook my head, and thanked her for her efforts; we parted friends. As I let myself out of her office, papers under my arm, I reflected that she obviously had missed the point of my joke—we're stuck in our problems because we're determined to solve them by using the very reasoning and processes that produced them in the first place. Like the donkeys, we must learn to appreciate the totality of things as well as our place in the total ecosystem.

Summary and Conclusions

To generate the power necessary to help maintain a healthy environment, individuals must think globally and act locally. This isn't possible unless, psychologically, people learn to cooperate with the Earth's metabolism. They must listen to the planet's wise, time-tested callings that coordinate the preservation of life. To receive global T-R messages requires scientific research that's balanced by our inborn senses, intuitions, and feelings. Ecological problems arise from the scientifically dogmatic environmental and academic communities' inability to recognize and culture people's global sensitivities and emotions.

Chapter Three

WHOSE BODY NATURE IS

Who cares if the seaweed deposited at the highest tide this month would have decomposed into a unique nourishing fertilizer to be carried to the waiting offshore waters at the highest tide next month?

"Okay, Dr. Doolittle, what are you going to do now?" I asked myself, as the lonely elevator silently descended from Susan Gould's office to the lobby of the building. "Of course, you can always listen to the nice literary agent, and hang it up..."

The doors opened into the dimly lit 54th Street exitway. A child sidled closer to her mother as I approached; the mother took a protective hold on the little girl's jacket.

I smiled at the little girl encouragingly. "Cleared up now, hasn't it?" I said to the mother. "Nice day."

The mother released her grip on the child and seemed to stand straighter as she smiled timidly and acknowledged that the sunlight pouring warm through the glass doors was an improvement over the rain of early morning.

I don't usually talk to strangers in the city; I was testing.

Yes, the woman did understand my words. Yes, she *was* less tense as I talked briefly about the weather. I bet she only would have withdrawn further if instead I had said, "Isn't it wonderful how the city is increasing the police force?" Even in the heart of Cementville, U.S.A., she responded visibly to the weather, to a natural phenomenon. I saw it with my own eyes. It affected her.

As I walked on through the glass doors, I carried with me the thought of that mother and her child. In the easy fantasy of a daydream, I took them and placed them in the natural world; I introduced them to the changing kaleidoscope of sun and rain, rivers and trees, mountain flowers and desert canyons, Indians and country people that I myself had discovered years ago.

I put them there with food and clothes, a small tent, inquiring minds, and a supportive community. I left them for fifteen years, day and night, winter and summer.

"If just one touch of rain and sunlight affected them today," I mused, "wouldn't a deep and lasting change occur over a span of fifteen years? Wouldn't the experience stimulate emotions and perceptions about nature that are foreign to that mother and child today?" I walked, lost in thought.

33

"Don't people feel differently sitting cross-legged in a field of sweet-smelling summer grasses than they do jolting through the city subway?" Turning east, I hiked toward Third Avenue through the towering, windowed canyons of New York City. Nature shook me out of my fantasy—feelings of anger and bitterness churned in my belly. What was she saying? Was she upset to see these hacked and piled pieces of her granite and to know that by many people their inanimate, stiff and unnatural lines were more valued than her canyonlands? Was she angry that the prejudice against her identified in my article was being denied, even as she struggled to exist here?

"Borscht," I murmured under my breath. If Mom or Dad or my high school guidance counselor knew that Nature was signalling to me, that I could hear her, that I was responding to her voice, I'd be in a psychiatrist's office before I reached Third Avenue.

I plotted my defense.

"No, Ma, Nature doesn't *talk* to me; we don't sit down and have a chat—she...uh...communicates, yeah, *communicates* with me. Yes, Mom, really. Look. Say you are sitting in a noisy cafeteria with music blaring and people shouting across the tables. Say you are trying to write an essay about your immediate thoughts and feelings. Now, be honest, Ma. Wouldn't your essay be differently written there than if you'd written it outside on a hot day with your back pillowed against a warm rock and the air singing with the lazy sound of crickets?"

I liked that; I figured I could even make some inroads with a shrink using an approach like that.

Encouraged, I continued the train of thought as I approached my left turn for uptown. "Mom, the only difference between the two pieces of writing is the environments they were written in! Nature is communicating with you in the second one. 'So what,' you say? You see, Ma, that's just the point. In our society, Nature and what she has to tell doesn't count for much. Remember when I was a kid? We lived in the country. Remember how I used to say 9 x 6 = 47, and Dad would correct me; sit me down and drill me? To me, 9 x 6 wasn't anything but boring. What did 9 x 6 have to do with me? Jeez, I had the rest of the 9 times table right; that's 90%. But that wasn't good enough; I had to be 100%. I wasn't allowed to play in the woods, or swim, or take the boat out on the lake, nothing, until 9 x 6 came out to be 54. Math—the meaningless manipulation of numbers—that was high priority. Nobody gave me academic incentive to realize the priority of Nature, to learn to treasure how different, how alive I felt when I was splashing around in that old boat, outside of four walls and a roof. So I grew up, Mom, and I never recognized Nature's influence. I never stopped hustling long enough to think about it and make the connection between listening to Nature and loving to feel alive."

I paused for breath. "Still with me, Ma? Okay. So I was all grown up, and in my work I was spending most of my days outdoors. Nights, too. I was sleeping on the ground, and it was influencing me. I didn't think about it; I liked it; I was happy to be there, so I did it. That was fifteen years ago, Mom. And do you know that it was eleven years before I pulled the evidence of my feelings together and recognized that Nature could and did communicate, and that what she had to say was equally important to 9 x 6 = 54?"

I let her mull that over for a minute, and then continued. "Now I'm really working hard, Ma. Now I'm trying to get people to recognize that they have been taught to ignore Nature and their natural feelings. I want people to see that they've been raised, like I was, to be prejudiced against Nature. Even you. And you remember how it felt to have the real estate man turn us down because you were a minority group member from the old country? You must remember those feelings. Well, Nature is being treated in the same lousy way, and if you listen hard enough, you can hear her complaining loud and clear. But we never were taught to listen to Nature, now were we? Instead, if we really did want to make contact with Nature, to be out in the woods by ourselves, we had to sneak out—almost as though we were having an illicit affair. It wasn't possible for us to have an upfront relationship with Nature."

"There," I thought smugly as I approached the glass doors of Audubon headquarters. "That should take care of Mom and Dad."

I was actually smiling when the remembered rasp of my old high school guidance counselor sounded through my head. "Come clean, Cohen," he said. "If you're so sane, how come you're fifty-two years old and having a conversation with your mother, who's been dead for thirty years?"

"Scram," I replied, as I stepped into the elevator and punched the button for the education department.

Chapter Four

A TIME TO BE BORN

Why do we deny and discourage life? What is so hateful about this place, that we are driven to kill it?

Reaching my office, I dropped into my chair and tossed the packet of papers I had been carrying onto the desk. Seeing the creamy buff of the clean manila envelope edging out under the little white heap of rejection notices, I recalled the second half of my article. I was rather relieved I had not submitted it to Susan Gould.

I smiled to myself, imagining her professional reactions, knowing that Part I was bland alongside the academic and scientific controversy Part II could ignite. I had prepared myself for its controversy; I had spent four days in a research library, studding the manuscript with footnotes, documenting every conceivable point of argument. This way, I figured, the article could be read as I originally wrote it, free of the weight of academic and psychological jargon; then optionally re-read, using the numbered references to locate documentation quoted at the end of the article.

Wondering what my perceptions of it would be now, I slit the envelope, drew out the article, and began to read.

§

PART II

Why must this happen? What insanity forces us to kill this bay for money and power? Why does a natural area appear so foreign that we can't relate to it, to life, or to each other?

Mental illness must be the source of our environmental problems; it is madness to murder one's own life-support system. I think about this as I float towards the ocean, because it is here that an answer lies. I can sense humanity as the canoe gently rocks in waves stirred by a local offshore breeze.

Various aspects of our inner relationships can be found in this bay and in Nature.[1] Our environmental problems remain and multiply because we are not conscious of their roots and symbols.[51] My mind wanders to the roots that make the symbols. The roots go way back, deep into the earth.

I'm not conscious of the feelings that shape them, but I know they're there.[2] They are a part of the birth process.

A human embryo's womb environment is supportive and feels euphoric to the unborn child. Birth is a difficult transition. Life outside the womb is challenging, especially in comparison with the static security of the womb. The roots of humanity are unconsciously formed in the beginning of human life.[3] These thoughts linger in my mind, and the remainder of the canoe trip is experienced within their framework.

I think to myself: if I were an eight month old human embryo curled in this canoe right now, I would not have the mental faculties to remember this canoe ride.[4] It would be unconscious.[5] A feeling level memory of this canoe journey might be recalled, for an embryonic brain is capable only of experiencing and recording feelings.[6]

If I were an embryo sitting in this canoe, I would be aware of feeling the warmth of this morning sun. My embryonic mind would record other feelings: the support of this water and its buoyant, rocking motion; the satisfactions of food, salinity, and excretory processes found here; the comfort of companionship and contact with other life that surrounds the canoe; the sounds of wind and water circulating; and my contentment as I sit securely snuggled and protected inside this ribbed frame.[7] [8] If I were an embryo in this canoe, I would know life as the euphoric harmonious sensations of this gentle bay.[9]

Even today, as adults, there lies in each of us the desire to experience the euphoria that I experienced in my own mother's womb.[10] This euphoria is not conscious, but one way I can recapture its peacefulness is by canoeing on this tranquil inland waterway.

At other times as well, euphoric womb feelings rise into our consciousness.[12] Their presence in myself and in others proves to me that at some level we carry memories of our euphoric relationship with our mother's womb.[11] There is no other explanation for the special comfort people derive from a water bed, a warm bath, the rhythmic heartbeat of music, the chemical euphoria of liquor and drugs, the peaceful coziness of the fetal position, or the *heavenly* emotionality of certain religious experiences. These comforts are womb memories that have been brought to consciousness.[13]

A person is the product of two mothers: *Mother* Nature, and his or her human mother. Mother Nature is the universal womb of life. Her life-giving relationships are duplicated in the womb of our human mothers.

Our womb feelings stem from contact with both of our mothers. They are psychological memories of life with Nature and life in our human mother's womb.

In the human womb, an infant experiences the oceanic bliss which can be achieved only through the harmony of Nature, and these feelings are re-experienced in an adult whenever there is a release of tension, for tension-release is an unconscious signal that a niche for survival in the womb of Nature is available. The womb is our embryonic experience with that niche; it is euphorically remembered as the ultimate niche.[3] This womb-euphoric state of mind is imprinted on our memories. At least subconsciously, we all yearn for a return to the womb—to be one with the harmony of Nature.

People also obtain euphoric feelings from material possessions.[14] I was mildly entranced when I bought this canoe. It was a deficiency need being satisfied instead of euphoria from harmony with Nature.

Far beyond that which is necessary for survival, we adults often crave the euphoria of material technology: money, larger shelters, transportation, recreation, protection, and power.[15] We seek additional womb euphoria feelings when we're not fully satisfied by our relations with people and Nature.[50] In our American culture, we are seduced to seek this additional euphoria in materialistic, technological sources. Advertising daily barrages our minds with the message that to be happy we must consume. What is the payoff? Many appeals are aimed at our insecurities; will we ever have all the money, power, or sex we need to be seen *by others* as being successful? If we were at peace with ourselves and could know the only real security—a basic feeling of oneness with self and Nature—then appeals to our insecurities would lose their power.

"That's a big part of our problem," I think to myself. "Our materialism puts pressure on the natural world."

This is a very special bay; it is part of the Bay of Fundy tidal system. Common here are eighteen to twenty-eight foot tides that retreat as far as two miles from the shoreline. What was a sky-blue lake full of water at 6:15 AM is an exposed, mile-long, clam-filled flat by noon.

A narrow neck separates this bay from the open ocean. Every six hours, like a dancing, snorting stallion, the bay charges in or out causing a turbulent waterfall that changes direction four times each day. It's a wondrous phenomenon to me; a dangerous irritation to the uninitiated boatman; and a source of potential tidal power profit to the exploiter.

Now my canoe enters this wild neck of water. It's like the birth experience itself; even though I'm an adult, the serenity of my inner-bay euphoria is severely interrupted.

The surging water guides the canoe, from within the bay, through the narrow neck, towards the ocean. The waters around me boil, weave, and swirl in turmoiled eddies; land flows by like the view from a runaway truck. The seething energy of the universe is alive, careening around the canoe, forcing it to spin, rock, jump in rhythm with the turbulence. A spill in the icy undertow could prove fatal.

If I were an embryo now, sitting in this canoe, I would be trauma-
tized.[16] The powerful feelings of this new experience would be emotionally fixed as helpless fright, and would be kept out of my consciousness
whenever possible. These raging narrows are similar to the birth canal; I
am borne into the open water of the ocean.[17]

Now I am outside the bay's protection.

There is a striking contrast between the serenity of the inner bay and
this ocean environment; the ocean's callings relate the difference. Survival
here asks that I cooperate with the challenges of wind, water, rocks, and
waves; that I make full room for the forces of Nature, for here I can easily
be overwhelmed. The challenges of the ocean are similar to the callings of
Nature to the newborn child, who is suddenly faced with hunger, thirst,
excretion, and respiration.*

Long ago I learned that a canoe in the ocean is more manageable when
another person is aboard to share the paddling, bailing, and fun. As
together we'd cope with the high winds and waves, I learned to appreciate
and enjoy the experience, and companionship, as much as I had enjoyed
the bay alone. Now I accept the ocean for what it is, and it does not baby
me. The ocean and I seek an equilibrium of power and reverence in
proportion to our respective needs. We share the wholeness of life.

If I were an embryo sitting in this canoe right now, my entrance into
the ocean world would begin a stressful period of readjustment from the
womb-like bay waters, just as Nature touches the newborn infant with the
tensions of self-preservation.[13] [19] [20] As an embryo, I would associate these
tense feelings of the wind and waves with the ocean and with being out
here, just as I would associate the calmness and serenity of the bay with
being in the bay.[22] [23] [24]

These are the same kinds of subconscious associations that a newborn
child makes: euphoria is associated with the womb, and tension and
discomfort with the outside and Nature.[25] [26] [27]

We develop prejudicial tendencies against Nature as, unconsciously
throughout life, our uncomfortable feelings about Nature's fluctuating
pulse don't measure up to our more static comfortable euphoric feelings
from the womb.[29] [30] Discrimination takes place as Nature is rejected and
womb euphoria is sought.[31]

The association rises into consciousness as language develops.[28] We
seek relief from Nature's tensions.[32] [33] [34] The technologies of adult life are
no longer sought solely because of their survival value; they are unconsciously desired for their relief value, their similarity to the euphoria of the
womb.[35]

*For further discussion of this point, see Chapter Six.

Perhaps now, one begins to understand why this bay must die, why we uncontrollably devour so many natural areas. As Americans, our relationship deficient hunger for womb euphoria is satisfied by the acquisition of technology. Excessive money, shelter, clothing, energy, and social acceptance provide satisfaction for those who are unable to find fulfillment in the pulse of people, Nature, or themselves.[36] While we gain technology, we lose the bay and its values.[37] The bay becomes merely a natural resource, grist for technology's mill.[32]

The canoe trip ends. It has illustrated that feelings from earliest childhood are transformed in our civilization into dependency on and infatuation with technology. It has clarified our subconscious fantasy, wherein technology returns us to the womb. Like a house, technology keeps Nature's tensions out and people in, inside the euphoric, womb-like environments provided by offices, cars, machines, and chemicals. We have learned that our Planet Blue (natural) feelings are dangerous, wrong, and banal because they fluctuate. To excess, we are taught that our materialistic culture is right and good, that it is our home.

Lame Deer:

Sometimes I think that even our pitiful tarpaper shacks are better than your luxury homes. Walking a hundred feet to the outhouse on a clear wintry night, through mud or snow, that's our small link with nature. Or, in the summer, in the back country leaving the door of the privy open, taking your time, listening to the humming of the insects, the sun warming your bones through the thin planks of wood; you don't even have that pleasure anymore.

You have changed women into housewives, truly fearful creatures. I was once invited into the home of such a one.

'Watch the ashes, don't smoke, you stain the curtains. Watch the goldfish bowl, don't breathe on the parakeet, don't lean your head against the wallpaper; your hair may be greasy. Don't spill liquor on that table; it has a delicate finish. You should have wiped your boots; the floor was just varnished. Don't, don't, don't...'

That is crazy. We weren't made to endure this. You live in prisons which you have built for yourselves, calling them 'homes,' offices, factories. We have a joke on the reservation: 'What is deprivation?' Answer: 'Being an upper-middle-class white kid living in a split-level suburban home with a color TV.'

I knew an old Indian at this time who was being forced to leave his tent and to go live in a new house. They told him that he would be more comfortable there and that they had to burn up his old tent because it was verminous and unsanitary. He looked thin and feeble, but he put up a terrific fight. They had a hard time dragging him. He was cursing them all the time. 'I don't want no son-of-a-bitch house. I don't want to live in a box. Throw out the goddam refrigerator, drink him up! Throw out that chair, saw off the damn legs, sit on the ground. Throw out that thing to piss in. I won't use it. Dump the son-of-a-bitch goldfish there. Kill the damn cow, eat him up. Tomorrow is another day. There's no tomorrow in this goddam box.'

The canoe trip has conveyed that we have become addicted to technology's euphoric power, and demand far more than we need. Our unceasing craving for additional money, possessions, and security fuels the exploitation of Nature's realm, demeans people to the status of consumers,[38] and is no less insane than suicide.[39] [40] [50]

We have imposed upon the environment to fulfill the often unthought-out fantasies of childhood.[42] [43] These fantasies are strongly ingrained; we often have rationalized our exploitation of the environment as a religious dictate. It is not. [44] It is prejudice for static euphoria and against Nature that has led to today's grim environmental projections.[21] [46] It is pollution from the subconscious mind.[45]

This canoe trip would not be complete without emphasizing one very special observation: the bay serves as both the womb and the adult environment for most of the creatures who live in it. To them, the gently fluctuating bay is both euphoria and Nature. The seal pup is weaned from its mother's womb to the bay; it is not conflicted by culture.[49] In turn, the seal pup and the other life forms of the bay do not exhibit signs of the exploitive self-destructiveness, strife, or mental illness that is common among people who do not fully recognize the congruency of the planet and the womb.

§

I read the article and then reread the notes I had collected to corroborate it. It occurred to me that the psychiatric concepts I had presented in the article were well-known, and were the science behind most effective forms of counseling. Yet something is missing from traditional psychiatry. As positive as has been the contribution of psychiatric concepts to people's relationship with people, they are derived from a culture that maintains a warped relationship with the planet. They have, in many ways, merely served to reinforce our prejudice against Nature.

Eric Fromm describes humanity's drive to destroy with force and violence as *wolflike;* a demeaning, inaccurate indictment of the wolf. The word *Nature* does not even appear in the *Psychiatric Dictionary.*

Thomas Harris, a well-known psychiatrist, suggests that the ultimate truth and moral order is that *persons* are important.[47]

"Where is the school of human development based upon people feeling good about sustaining the natural system in order to sustain themselves?" I asked myself, shaking my head. "An important relationship between people and Nature exists, and will explode, if not faced squarely. We are being asked to become well-adjusted to an insane society."

REFERENCES

[1]Throughout history, one impression of human nature has been consistent: that man has a multiple nature. It has been expressed mythologically, philosophically, and religiously. (HARRIS R8, p1—[Resource 8, Page 1])

[2]The brain functions as a high-fidelity tape recorder, putting on tape, as it were, from time of birth every experience, possibly even before birth. (HARRIS R8, p9)

[3]The validity and reality of the uroboros symbol (womb feeling symbol) rest on a collective basis. It corresponds to an evolutionary stage which can be *recollected* in the psychic structure of every human being. It functions as a transpersonal factor that was there as a psychic stage of being before the formation of an ego. Moreover, its reality is re-experienced in every childhood, and the child's personal experience of this pre-ego stage re-traces the old track trodden by humanity.

All mythology says over and over again that this womb is an image, the woman's womb being only a partial aspect of the primordial symbol of the place of origin from whence we come. This primordial symbol means many things at once: it is not just one content or part of the body, but a plurality, a world or cosmic region where many contents hide and have their essential abode.

Compared with this maternal uroboros, human consciousness feels itself embryonic, for the ego feels fully contained in this primordial symbol. (NEUMAN R11)

The collective unconsciousness is the deepest level of the unconscious, containing psychic elements common to mankind in general and derived from phylogenic experience. (KARL JUNG *in* Nicholi, R19, p127)

The earth is as mother, the womb from which all living things are born, and to which all return at death, was perhaps the earliest representation of the divine in post historic religions. (DENA JUSTIN *Natural History*)

[4]The mind and soul of the sleeping newborn are not yet functioning. The child being in a state similar to his mother's womb—without consciousness. (ZUR LINDEN R12, p10)

[5]In the newborn there already exists a *condition* that is at least similar to consciousness and from which later consciousness evolves gradually

increasing in content—otherwise the newborn would not be able to utilize his experiences. (STIRNIMANN R12, p14)

⁶The centers and connections of the cerebral hemispheres are already partially matured at birth and therefore exert a certain influence on the behavior of the child. The psyche attempts to express itself from the beginning... the newborn... crying indicates a desire to be released from a disagreeable situation. (REMPLEIN R12, p14)

⁷The newborn already possesses an emotionality on the basis of which he produces signals. (AENGENENDT R12, p14)

⁸During this nine months there occurred a beginning of life in the most perfect environment the human individual may ever experience. (HARRIS R8, p40)

⁹Much early learning occurs without identifiable reinforcement as in *imprinting* with animals. Infants in the first week of life apparently have elaborate built-in mechanisms for initiating some distinctive forms of apprehension. (GRENELL R6, p77)

¹⁰Even in the mature adult, the longing for the protective orbit of the mother's womb—the situation as it once existed, never ceases completely. (FROMM R5A, p43)

¹¹A problem with many premature babies is that in later life they do not appreciate or respond to physical contact with people. The conditioned response does not attain full strength because early detachment from the womb and extended periods in an incubator prevent normal physical association with the womb. (BLANCHARD verbal)

¹²Recollection evoked from the temporal cortex retains the detailed character of the original experience. When it is thus introduced into consciousness, the experience seems to be in the present. Only when it is over can it be reorganized as a vivid memory of the past. (PENFIELD R8, p9)

¹³Infants on the first day of life can learn; indeed, learning in the period before and during language development is probably better than at any age. (GRENELL R6, p76)

¹⁴Recorded experiences and feelings associated with them are available in as vivid a form as when they happened. These experiences cannot only be recalled, but relived. (HARRIS R8, p11)

[15]Inferiority as a universal human feeling originates in the child's feelings of inferiority because of his smallness and helplessness and that inferiority is countered by the need to seek a dominating or superior position. (ALFRED ADLER in Nicholi, R19, p125)

[16]The feelings produced by birth trauma were recorded and reside in some form in the brain. This assumption is supported by the great number of dreams of the *drainage type* variety that people experience following situations of extreme stress. (HARRIS R8, p40)

[17]Being born is such an anxiety-producing experience for the infant that it is the prototype of all later anxiety. (RANK R10, p174) (FREUD R8, p41)

[18]At birth, the infant is flooded with overwhelming unpleasant stimulations, and the feelings resulting in the child are, according to Freud, the model for all later anxiety. The infant is *rescued* by another human being who begins the comforting act of stroking. This is the first incoming data that life *out there* isn't all bad. It is a reconciliation, a reinstatement of closeness. It turns on his will to live. Deprived of it, physical death will occur. (HANK R8, p41)

[19]We should remember that a baby is afraid of the world and is probably yearning for a place to which no one can ever return. (BERNE R1, p95)

[20]To the newborn, occasions of danger may call out terror... extrauterine pressure on the lungs calls out what we would, in later life, call rage. The screaming and kicking infant may be experiencing oxygen starvation. (SULLIVAN R16, p49)

[21]Cultural symbols have been used to express "eternal truths." They can evoke a deep emotional response in some individuals, and this psychic change can make them function in much the same way as prejudices. (JUNG R23, p93)

[22]A new experience is somehow immediately classified together with records of former similar experiences so that judgment of differences or similarities is possible. (PENFIELD R8, p10)

[23]Initially, events, or their internal representations—ideas—are brought together solely on the basis of having occurred contiguously in the experience of the subject. Each undifferentiated contiguity gives rise to mediated contiguity. Mediation tremendously enlarges the scope of associative pro-

cesses in that it allows for connections between two events even if these events have never, and never will, occur in temporal or spatial contiguity. (POLLIO R13, p79)

[24]The effect of an emotionally disturbing event may be multiplied as a consequence of the important effects of various antecedent vicissitudes of early life, which come to serve actually or symbolically as analogous prototypes. Every emotional conflict has its earlier antecedent. (LAUGHLIN R22, p29)

[25]Lipsett and his workers have presented evidence for responses involving mouth stimulation and head turning responses in the neonath through the use of reinforcement contingencies. (WOLLMAN R17, p74)

[26]The baby can be conditioned during the last two months of pregnancy. (The Conditioning of the Human Fetus in Utero, *Journal of Experimental Psychology,* 1948, 338-346. SMART R20)

[27]Association refers particularly to the relationship between the conscious and unconscious, ideas in the former being conducted to the latter. (HINSIE R9, p71)

[28]We know that the neurotic process arises with the development of the capacity to act and think, and ultimately speak in symbols. The symbols of neuroses parallel the symbols of language. The neurotic distortion of these symbolic functions occur as a result of a dichotomy between conscious and unconscious processes which starts early in the development of each human infant. (KUBIE *in* Mowrer, R21, p22)

[29]For a little child, it is safer to believe a lie than to believe his own eyes and ears. (HARRIS R8, p36)

[30]It is possible to make one certain deduction: in the earliest months, a central emotional position is frequently established in which the individual will tend to return automatically for the rest of his days. Establishing a central emotional position may be one of the earliest among universals in the human neurotic process, beginning in the presymbolic days of infancy. Whenever the central emotional position is painful, the individual may spend his whole life defending himself against it using conscious, preconscious, and unconscious devices whose aim it is to avoid this pain-filled central position. (KUBIE *in* Harris, R8, p42)

[31]The first psychoanalytic views were based upon the conflict between instinct and the outer world. The most pregnant formulation of all analytic psychology is that the psychic process reveals itself as the result of the conflict between instinctual demand and the external frustration of this demand. (REICH R14, p287)

[32]The will to power is the guiding force, a masculine characteristic. The need to subjugate the mother (Mother Nature? Mother Earth?—*MJC*) and compete successfully with the father, the will to power, is the guiding force in human behavior. (NICHOLI R19, p125)

[33]The suckling child's world is a changeable place where *anything can happen* and where terrifying things do happen. Somewhere is something which is warm and loving and makes him feel secure. It also allays his hunger and strokes his skin so that he falls into a refreshing sleep. His greatest security comes from being close to this warm and living influence. (BERNE R1, p96)

[34]We say a person is prejudiced against Jews or Blacks and mean that he harbors antagonistic feelings toward them. Prejudice also means preconceived opinion. (GUTTMAN R6, p445)

[35]Organization of ideas contributes to the structuring of stable images relating to things. Such images finally become so structured that even if objects only roughly and erroneously resemble already well developed images, the objects may nevertheless be perceptually, incorrectly interpreted as conforming. The remembered images misconstruct ongoing sensory experiences and show great resistance to being corrected. (GRENELL R6, p19B)

[36]The pleasure-pain principle is a regulatory mechanism of mental life whose function is to reduce psychic tension that has arisen as a result of drives pressing for discharge. It tries to undo the effects of disturbing stimuli in a way that will most easily provide satisfaction. (HINSIE R9, p275)

[37]The most prosperous countries in the world show the most severe symptoms of mental illness. (FROMM R5A, p19)

[38]The self-attitudes learned early in life are carried forever by the individual, with some allowance for the influence of extraordinary circumstances. (SULLIVAN R8, p44)

[39]Only the rare individual has an accurate idea of what his drives and needs can make him do without even being recognized. It is in this last respect that the greatest and most frequent errors are made. (BERNE R1, p65)

[40]People think they are looking for security, but what they are really looking for is a feeling of security, for actual security, of course, does not exist. The feeling of security is increased by having means available for the relief of tensions and anxieties and the gratification of wishes.... Seeking freedom from anxiety does not always correspond to what we commonly think of as *seeking a safe situation.* (BERNE R1, p59)

[41]Modern man does not understand how much his "rationalism" (which has destroyed his capacity to respond to the spell of symbols and ideas) has put him at the mercy of the psychic underworld. (JUNG R23, p94)

[42]We have some objective evidence from recent experiments measuring the brain's electrical responses (evoked potentials) to varous kinds of subliminal (beneath conscious) messages. The experiments suggest that the brain is fairly humming with unconscious thoughts and emotions that shape what we pay attention to and what we tend to repress—to use the psychological term. (SHEVRIN R15)

[43]Understanding the *age three* fear is essential to freeing the Adult for processing new data. This is the fear—the archaic fear of the all-powerful Parent—which makes persons *prejudge* or which makes them prejudiced. A person who is prejudiced is a person who accepts inaccurate concepts as ultimate truths. He is afraid to do otherwise. This produces the contamination of the Adult, and this contamination allows prejudice, or unexamined Parent data, to be externalized as true. (HARRIS R8, p57)

[44]Psychoanalysis is a dynamic conception which reduces mental life to an interplay of reciprocally urging and checking forces. (HARRIS R8, p2)

[45]In the act of destruction of life, man sets himself above life, he transcends himself as a creature. (FROMM R5A, p42)

[46] (Although they can be prejudicial) Cultural symbols are important constituents of our mental make-up and vital forces in the building up of a human society. (JUNG R23, p93)

[47](HARRIS R8, p220)

[48]Although early experiences which culminated in the position cannot be erased, I believe the early positions can be changed. What was once decided can be undecided. (HARRIS R8, p43)

[49]In the more olden times, the world of nature and of human relationships expanded in a rather orderly manner, keeping pace with the maturity of the child. There was free space around his home, a field, a meadow, an orchard. The apartment child, and to some extent even the suburban child today has been greatly deprived of his former human and infrahuman companions. (GESELL R8, p147)

[50]An adult's problem is to find out the best way to handle other energy systems in order to gratify his wishes speedily with the least danger. (BERNE R1, p64)

[51]Recollections can be more accurately described as a reliving than a recalling. Much of what we relive we cannot remember. (HARRIS R8, p7)

Resources

1. BERNE, ERIC. *A Layman's Guide to Psychiatry and Psychoanalysis.* New York: Simon & Shuster, 1968.

2. COHEN, MICHAEL J. The Environment As Educator, *Environmental Education Report,* 1977.

3. FALKNER, FRANK, ed.*Human Development.* W. B. Saunders & Co., 1966.

4. FREEDMAN, ET AL.*Comprehensive Textbook of Psychiatry.* Williams & Wilkins Co., 1967.

5. FREUD, SIGMUND. *The Problem of Anxiety.* New York: Norton, 1963.

5A FROMM, ERICH. *The Sane Society.* HR & W, 1955.

5B FROMM, ERICH. *The Heart of Man.* Harper & Row, 1968.

6. GRENELL, ROBERT & SABIT GABAY, eds.*Biological Foundations of Psychiatry.* Vol. 1. New York: Raven Press, 1976.

7. GUTTMAN. *Psychoanalysis, Observation and Theory.* New York: International University Press.

8. HARRIS, DR. THOMAS. *I'm O.K., You're O.K.* Harper & Row, 1967.
9. HINSIE, LELAND & ROBERT J. CAMPBELL. *Psychiatric Dictionary.* Oxford University Press, 1960.
10. NEEL, ANN. *Theories of Psychology.* Cambridge, Shenkman Publishing.
11. NEUMAN, ERICH. *The Origin and History of Consciousness.* Princeton University Press, 1970.
12. PEIPER, ALBRECHT. *Cerebral Function In Infancy and Childhood.* New York: Consultants Bureau, 1963.
13. POLLIO, HOWARD R. *The Psychology of Symbolic Activity.* A-W, 1974
14. REICH, WILHELM. *Character Analysis.* Simon & Shuster, 1972.
15. SHEVRIN. Glimpses of the Unconscious. *Psychology Today.* April, 1980
16. SULLIVAN, HARRY STACK. *The Interpersonal Theory of Psychiatry.* Norton, 1953.
17. WOLMAN, BENJAMIN. *Handbook of General Psychology.* Prentice Hall, 1973.
18. NASH, RODERICK. *Wilderness and the American Mind.* Yale University Press, 1978.
19. NICHOLI. *Harvard Guide to Modern Psychiatry.* Harvard University Press, 1978.
20. SMART & SMART. *Child Development and Relationships.* New York: Macmillan, 1972.
21. MOWRER. *Psychotherapy Research.* Ronald, 1953.
22. LAUGHLIN, HENRY. *The Neuroses.* Butterworths, 1967.
23. JUNG, CARL. *Man and His Symbols.* Doubleday and Co. 1964.

Summary and Conclusions

The evolution of prejudice against Nature tendencies is observable in the emotional growth of a person. On a feeling level, people clearly remember their chemically supported life in their mother's womb, their birth, and the first two years after their birthday. The events of this early period are only remembered as feelings, for in the early stages of development the conscious mind is too immature to record memories as events. Psychologists have found that subconsciously not only does an infant retain its euphoric womb feelings, its "security," but it tenaciously seeks these feelings during the remainder of its life.

At birth, the helpless infant begins the stressful period of adjustment from the womb to the more disturbing forces that affect survival in Nature; suffocation, hunger, thirst, a niche, temperature adjustment, and loneliness. Like womb euphoria these uncomfortable feelings are also stored in memory. They are later symbolized and associated with the *outside* and with the realm of *Nature*. They are called *Nature*, because they aren't created by people or culture. Throughout life, people's disturbing feelings about Nature's wild fluctuations don't measure up to their euphoric womb memories. Unjustifiably, nature is rejected while "security" is sought. We are acculturated to our infatuation with technology because like a house, technology keeps nature out and people in protected womblike environments, such as offices, cars, machines and chemicals.

People become addicted to technology's euphoric power, demanding far more than they need for harmonious survival. Our craving for additional money, material possessions, and security fuels the socio-economic juggernaut that exploits Nature's "raw materials," demeaningly makes people into "consumers," and is no less insane than suicide for it injures our global and personal life support systems.

Chapter Five

IN TIME RETURNS THE RIVER

Am I the only one who recognizes that human consciousness is shared, at some level, by this community from which we gained all of our life processes?

By the time I left the Institute office that afternoon, I felt that I was being dissected and so was Mother Nature. I had unhealthfully become accustomed to the discomforts of being taken apart; but Nature, finding it intolerable, was signaling for my attention. Let me explain what I mean when I say Nature was "signaling" or when, in her office, I told Susan Gould that, "Nature tells me something different."

Perhaps the closest that our culturally educated and conditioned minds come to being conscious of Nature is found in the latter portion of the statement, "Excuse me, I hear *the call of Nature*." In this clause, I believe that we correctly symbolize and convey our feelings of excretion to be an expression of Nature; the natural tension of the excretory process leading to the released feelings that follow defecation or urination.

What is Nature? We seldom ask this question because we take for granted that we already know the answer to it. But in my twenty years of formal education and in the quarter century that has followed it, I have observed that there are many differing definitions of Nature. I have yet to find a definition that is more universally accurate than Nature being a harmonious process of tension building and release (T-R) between the energies and entities of the planet and cosmos; an irregular, disorganized, pulsating process that is found in people, plants, animals, climate, rocks, seasons, the planet, and the solar system. Nature is the fluctuations of the tides, feelings, and populations, the communication between, and/or mutual consciousness of, individual life forces, the seldom recognized but omnipresent heartbeat and dance of the universe.

As best I can determine, Nature's T-R process is *experienced* as a *sensation* that pervades all matter, that guides every entity into fluctuating but self-sustaining harmonious relationships with other entities. Nature appears to be a relationship that enjoys and promotes life. But in the same way that I cannot fully describe the color *green* by using the *black* ink of this typeface, words cannot fully describe Nature because Nature is an active experience, the life experience. One must *experience* the color green in order to know it, and just as green changes, fades, or intensifies with time, so Nature fluctuates from moment to moment or year to year.

53

As it does with everything else that it recognizes, our language labels Nature. The labels are our static words and symbols that often seek consistency in Nature. They are frequently inaccurate because Nature is in constant flux. Even death does not appear to be static.

Somehow, we must construct fluid, congruent symbols and ways of relating, if we are to become truly conscious of Nature. We can't fully relate to a blue planet if the glasses that we wear let only red light pass through them. We can't fully comprehend the motion of Nature, if our symbols tend to perceive still pictures. We can't measure the life of a wolf or a bog with a yardstick.

Nature has been most of my daily environment. To better understand Nature, I have chosen to live close to her. During this period, I have not been conditioned or rigidly biased by human words or other human perceptions and values, for they are often absent in natural environments. Instead, I have sensed Nature on her own terms in her own home. I have discovered some of the incongruities between how my culture taught me to experience Nature and what Nature has to say for herself.

I have come away with the notion that Nature's fluctuating T-R sensations are experienced by people as feelings, Planet Blue feelings. I am convinced that the natural T-R feelings that people experience are in reality a voice of Nature, a vital part of each person's self-preservation, instinct, and biology. That our natural feelings are acculturated and are given little value in our society is a phenomenon that sorely needs examination, for that practice silences global thinking, the voice of Nature, the voices of self-preservation of and on the planet.

I have discovered that unknowingly, my childhood and formal education taught me to be prejudiced against my own natural feelings, while my years with Nature and different subcultures have taught me to respect those harmonious Planet-Blue sensations in myself and others. I have learned to differentiate natural feelings from their conditioning by society. My natural feelings are the stimulus that I am describing when I say that, "Nature was signaling me for my attention." I have only recently become conscious of them and integrated them, for their pulsing wholeness was educated out of me by my former surroundings and schooling.

My feelings told me that Susan Gould and the editors were trying to divide fifteen years of my life into pieces of their expertise. Science? Psychology? Education? These labels were comfortable pigeonholes: familiar, static, euphoric. My life experiences were whole. They had brought me to consciousness that enabled me to perceive the tidal bay's dilemma and the source of that dilemma.

My experiences might be worthless to publishers, but they are invaluable to me. To me they are not merely a composite of ideas, they are a life. I have had my fill of being subdivided into the conceptual schemes and expectations of others. Nature has taught me to be Planet Blue and to treasure the wholeness of my experiences.

§

Quite frankly, I never originally intended to write this book. I just wanted to place an article that would help to explain why people's relationship with Nature was so destructive; to describe the subconscious womb-euphoric factors that are being overlooked while our environmental problems increase; and to convey that our relationship with Nature is based on prejudice and might more constructively be approached with that fact in mind.

In time, I became greatly aggravated by the shortsighted, prejudiced against Nature responses that my words and observations had drawn from portions of the academic and publishing worlds. Their reactions made me recognize just how much Nature had taught me over the years about herself and the integrity and extent of the life process.

Contrary to the beliefs of most of the publishing and academic worlds, the planet is undoubtedly a living being, a life system whose wholeness sustains life and can be injured as you or I might be, if we were poisoned, stabbed, or raped. The planet is as susceptible to the hurtful effects of prejudice, as is any person belonging to a persecuted minority, sex, race, or religious group.

I'll never know whether it was Nature's urging or academia's denials that drove me to write the remaining chapters. Perhaps it was the play within myself of one upon the other. I do know that about a month after I met Susan Gould, I decided that academia and academicians exist and have assumed the right to divide Nature and me into disconnected bodies of knowledge. Subject by subject, they teach what they believe they know about Nature, people, and health.

But the planet Earth also exists and is knowledgeable. It is the wholeness of its intact natural systems, and it deserves the right to teach Nature's wholeness to us.

I believe that we all would be repulsed as I have been, if we knew the tragic effects of our American lifestyle upon living systems of the Earth.

Educators have been taught to subdivide Nature, to shatter her wholeness and discuss the shards within the sterility of a classroom. Unwittingly, our civilization has been dedicated to removing the individual from the natural world. We examine the planet with our emotionless and rational interest, not in mere intellectual curiosity, but to discover new sources of culture-induced euphoria and profits.

Most educators have focused on utilizing language to understand nature, but nature is a feelingful process, we learn to understand the words but don't learn the music. The double message inherent in their methods of teaching and their chalk-dust surroundings remind me of the hypocrisy of a college health education class, where the subject was the health hazards of smoking, but where the teacher and most of the students smoked. Can harmony with Nature ever be achieved by those who, inside the controlled monotony of four walls and a roof, try to make Nature conform to their static laws?

I have witnessed first-hand that while academia has purported to be working at rectifying the problems in our relationship with Nature, the relationship has deteriorated steadily. Today, the hopes and aspirations of the mass culture are more divided, more institutionally dependent, and more distant from and demanding of Nature than they were two decades ago. It is easy to see that the twin systems of American education and the American media are serving well the exploitive industrial, corporate, and political sources that help finance their existence.

Since leaving college and finishing my formal education, I have learned from Nature that when you place a straight line on a large globe, it may come out straight—but when it does, it places stress on the globe and produces circular social problems. Vicious circles are caused by the misplaced energy of the roundness of the planet expressing itself in the lives of people who are intent on straightening out the planet.

A museum curator in Bear Butte, South Dakota, explained the dynamics of a circular problem to the Expedition students. He told us that one hundred years ago, when the immense buffalo herds spanned the plains, there was more meat on the hoof than there is today, despite our modern practice of cattle farming. He said that if we were to bring back the buffalo herds that were once here, we would find we could not keep them alive on the plains because we have destroyed the prairies—the incredibly nourishing plant communities that were needed to sustain the herds. First, we would have to bring back the proper plants, the curator explained, but we can't do that because those plants are dependent upon large herds of buffalo for soil aeration and fertilization.

We asked if there were no solution to this circular problem, which had been caused by straight line intentions for the transcontinental railroad and the elimination of the Plains Indians.

"Not a simple one, no," he replied. "If a solution were to be found it would have to be whole. It would have to establish a *cultural* relationship between people and Nature that would allow Nature to take its complete, slow-but-sure course of building up the soil, the plants, and the buffalo herd together."

That solution would be a holistic, wisely civilized approach that is missing in our historic, impatient, *we can conquer Nature* straight-line civilization.

There are two ways for people to attempt to perceive the cyclic wholeness of the planet. One way has been to examine our environment. Oddly enough, taking this route may have been humanity's greatest mistake. Our grave error is that by dividing the planet into humanity on the one hand and the environment on the other, we have divided the whole. It is for this reason that we never have been able to find wholeness—we have denied ourselves our place within the cycles and being of the planet. Thus the more we dissect the workings of our environment, searching for static euphoric answers, the more we divide and destroy it.

People seem to be the only organism that relates to the living Earth by dividing it and trying to reduce it to a scheme of unnatural physical laws. Perhaps no other organism has had the ability or the audacity to do so. In turn, no other organisms are living disharmoniously with the planet. All others have established a cooperative equilibrium with the earth, adapting to the fluctuations of stress between themselves, with their community, and with the planet.

There is another way to experience the wholeness of the planet: it is to consider the wholeness of a person. In a most vital and important way, a living person and the living planet are exactly the the same thing—they are life. As with systems, what pertains to one pertains to the other. Our historical ways of understanding Nature have ignored the congruence of people with Nature.

We have euphorically but falsely perceived the planet to be a jigsaw puzzle. We have blindly related to it piece by piece. But we are an integral part of the wholeness of the planet, and each time that we have cut out and related to an isolated bit of Nature, we have proportionally lost some of our identity as whole natural beings. We have replaced our own wholeness with the very attributes we have assigned to the planet and to Nature. It is worth considering the full social impact of our fragmented approach to life:

- As we have made the planet into laws of Nature, we have in turn become subjects of these laws instead of fully free organisms.
- As Nature has come to be expressed as mathematical equations, we have become *mathematicians* instead of whole people.
- As the planet has come to be seen as natural resources, we have become resourceful exploiters of Nature, instead of cooperative members of a supportive Earth community.

- As Nature has become vulnerable to the objectivity of science, we have lost the value of our feelings.
- As Nature has succumbed to our power, we have become subservient to those in power.
- As we have interrupted the community of Nature, we have lost our sense of community; as we have become alienated from Nature, we have become alienated from each other.
- As we have capitalized on Nature, we have become capitalists rather than human.
- As we have competed for Nature's *raw materials,* we have learned to relate competitively. As we have treated Nature as grist for our civilization, we have learned to treat each other uncivilly.
- As we have identified ourselves as masters of the planet, we have lost our sense of place.

We, too, can share in the natural euphoria of the Earth's harmony, but only when we relate to the planet as an integrity that is healthy when it's whole, only when we recognize ourselves as a whole system and part of the planet's wholeness, and only when we personally refuse to be divided. If we do not adapt ourselves and our culture to the wholeness of the planet, the planet will maintain its wholeness by eliminating us. The planet's refusal to recycle our pollutants and impact is a sign of the shortsightedness of our view.

For the past decade, I have been a part of an effort to reverse the destructive trends of our culture through an accredited educational process that I have described in my previous books. We have created an expedition community that encourages each person to relate to him/herself, and to the Earth, holistically. In this community, every individual is recognized as a subjective, concerned, cultural, and natural being, who can learn to see him/herself as a harmonious facet of the planet's pulsating equilibrium. My students learn to acculturate themselves and professionally teach others to acculturate to the vast organism we call Earth. With this expedition *family,* many of the social problems which plague America have subsided, as our community participants have learned from the planet to enjoy life by skillfully and healthfully maintaining it.

In school, I once taught students that the ingredients necessary to maintain life were sun, soil, air, and water. Now, living directly with Nature has shown me and my students that although these ingredients are still present, we have poisoned them to the extent that life is on the decline in many areas. Obviously, the list which I was given as a student was incomplete.

The ingredients that I now believe necessary for life are: a sense of wholeness, an acute awareness of our self-preservation feelings, recognition of the planet Earth as a living organism like ourselves (which includes earthly properties of sun, soil, air, and water), and an awareness of human discriminatory practices against Nature. When and where these four factors are found, life, including human life, is presently flourishing in a living equilibrium with the planet. The Expedition Institute's goal is to instill these biologically sound values in every participant.

The texture and fabric of a living person are exactly the same as the living structure and process of the planet, and of any other living organism. I now believe that to attain a proper comprehension of the wholeness of life and the planet, people not only should study the environment first-hand and their culture's impact on it, but should study their natural selves and the impact that their culture has made upon them. Every individual should recognize and investigate the layered coats of education and cultural conditioning which conceal his or her Planet Blueness.

Although different, this approach has proven to bridge innumerable gaps that have plagued humanity's relationship with Nature.

By examining myself, I can perceive my own blueness. The truth is that I can better understand my own experiences and life than any other individual can understand them. I am the most direct connection between my sensing and perceiving organs. I can directly perceive and experience the wholeness of the life process, because I am alive and can conceptualize.

This is an entirely different learning process than that followed by academia.

Within the confines of academic thinking, to perceive the wholeness of life and the planet, I must first accept the limited, predigested sensations fed to me by the indoor environment of the home or classroom, and the bias of the instructor. I must remove my feelings from all interactions in order to objectively interpret them. I must divide, classify, and label my observations and perceptions, convey these to others for evaluation and correction, and receive back an acculturated message. This message I must then learn to experience as an accurate description of life and the living planet.

This academic process is divisive and remote. It denies the value of personal thoughts and feelings. It is subject to gross inaccuracy because the chain of communication from life to the symbolizing of life is lengthy, and thus is more liable to weak links. Could we have permitted and condoned the atom bomb, if we had not been conditioned by academia to accept the *objectivity* of science as more valid than the truth of our emotions?

Our mass culture often prospers by teaching us to be out of step with the wholeness of Nature. We learn to unholistically organize our observations of life into logical, rational, and reasonable laws and sequences. Presently in vogue are notions that Nature operates logically through physical laws of Nature; that the planet is composed largely of lifeless inorganic matter and is feelingless and without consciousness; that people's feelings are irrational in comparison with the purity of scientific facts; and that human energy and emotion interferes with the understanding and proper use of Nature. These observations may be culturally understandable to us, but their application has proven to interrupt the process of life on our planet.

Life pulsates to its self-imposed, irregular rhythm, while we attempt to impose order where none may exist. Life, and therefore the living planet, is disorderly: a series of vibrating, diverse, tension-producing and tension-releasing relationships that are frequently functions of specific times, places, components, properties, and cycles. They seldom again happen upon each other in exactly in the same way.

Our civilization has emotionally and historically perceived the disorganization, diversity, and interchangable pulse of Nature as annoying, almost intolerable. After our first steps toward cultural euphoria and away from the random fluctuating wholeness of the planet, we proceeded to prejudicially bulwark ourselves from Nature's lively variety by hiding in the "civilized" fortresses of four walls.

But living with Nature, I have learned a different lesson. Nature has taught me to be a civilized person by being whole, by respecting and cherishing her eons of life experience and by attempting to bend to her life-giving fluctuations on the Earth when it is possible.

Summary and Conclusions

An enigma, Nature's "order" is an irregular fluctuating tension buildup and release process (T-R) between the entities of the universe. In people, T-R is experienced as natural feelings, such as thirst, hunger, excretion and niche support. These serve as signals from the global life system; they're Nature's guiding voice for survival.

Geologically, the planet Earth is an intact whole life system, as is a human being. As interdependent life systems, people and the planet are congruent; what happens to one happens to the other. To subdivide a life system is to diminish its vitality. As we dissect the planet or a person in search of excessive security euphorias (Chap. 4), we disturb or destroy the former's ability to support life or each other. To gain life's wholeness and human preservation, people, by recognizing the value of their natural T-R feelings can recognize the unity of the planet and themselves as well as their culture's impact upon both. We must re-acculturate ourselves to enjoy and flow with Nature's life-giving fluctuations for they are the essence of life.

Chapter Six

TO GIVE LIGHT UPON THE EARTH

How convenient for us to conceive mud, water, and stones to be dead; to decide that other life has no consciousness, pain, or equality. What an incredible alibi we have created to soothe our guilt of killing for a profit.

"People can't have relationships with rocks, or deserts, or mountains," Susan Gould had told me. "The landscape consists of inanimate objects. Inanimate basically means dead; dead things don't interact; so your article about relating to them just doesn't make sense. Animals, yes. They're alive; some people do have relationships with them. But we can't relate to a planet; we don't relate to Nature, either prejudicially or positively."

"But the planet is just as alive—" I had started to say, then stopped. I felt foolish telling Susan that I knew the planet was alive because Nature had told me so. I said nothing and the subject was changed.

I remembered the conversation now, as I sat working in the Expedition office. I recalled that when I was eight years old my pet turtle had died. I had caught it in a lake the previous summer, and it had lived in our New York City bathtub until December. My mother threw it away in the garbage, for burial was impossible in the frozen ground. Three years later, I learned with dismay that wild turtles hibernate in the winter and can appear to be dead. I was horrified that I might have killed a living thing that I loved.

Susan Gould reminded me of the turtle because after fifteen years of direct observation it has become obvious to me that the planet Earth is as alive as you or me, perhaps more so. The evidence for its life is visable to anyone who values his or her thoughts, feelings, and personal experiences. The Earth appears to exhibit all the features of life that identify life, including consciousness.

I realize that to most people of my culture, my Planet Blue outlook makes me sound like I speak a different language. I felt like a person speaking Chinese in Georgia when I began to tell Susan Gould that I knew the planet to be alive because Nature told me so. I anticipated her reaction of baffled impatience. I guessed she would hear what I was saying as similar to mythology or metaphysics, and in our scientific civilization, they are practically taboo.

But Nature does not speak only in rational or scientific terms. She most often speaks through feelings, through the emotionality of tension and the release of tension through the inner T-R sensations evoked from the effects of the myriad of natural forces—the sun, moon, gravity, electromagnetic fields—which collectively comprise Nature. But our culture teaches us that emotion is usually unreasonable and irrational.

In order to relate the messages Nature has given me, I must speak about and with feelings. It is perfectly reasonable to me to do so, because feelings are Nature's language. But how can I communicate Nature's teachings in a manner that is acceptable to a scientific civilization when in order for me to be rational, I must speak about and with feelings—I must speak irrationally? This paradox is real. It is the major cause of America's separation from Nature, because blocked unconscious euphoric feelings are a prime source of our prejudice against Nature.

Let me begin with logic. Nature is unquestionably the universe— which includes the planet Earth. If Nature is alive, then chances are that the planet is alive—just as, if a person is alive, his heart is also alive.

Probably, we believe the planet to be dead because psychologically we killed our image of it when we first divorced ourselves from Nature and thus divided the planet's wholeness. We had the power to subdivide the planet, and we discovered that our imagery, not Nature, produced reality. We were subconsciously aware then, as we are now, that a divided, unwhole organism dies, for that is the way of Nature. We lit a flame with tools and technology, a flame whose glow tainted all that it touched.

For many people, the concept that the planet is alive is as inflammatory as was the pre-Columbian thesis that the world was round, not flat. I have heard that people were hanged for that assertion in olden times.

§

I have learned from Nature to feel Nature. I have experienced her by learning to relish the many sensations and processes associated with living outside. In doing so, I have discovered that I exhibit every life process that the planet exhibits. I conclude that I and the planet are alive.

What are some of the recognized properties of cellular life? Metabolism, temperature, chemical norms, reproduction, respiration, motion, reactivity, excretion, interaction, metamorphosis, energy, sensitivity, irritability, interrelationships, and self-preservation consciousness at some level? Look hard and you'll find that each of these functions exists in a rock, a mountain, a river, the land, and the planet. Each of them, in their own sweet time and way, go about the life process. It is not our way or velocity of life, but it is a way of life. Most important, it is life.

Without the physical or psychological containment of four walls and a ceiling, I have found that Nature daily leads me to observe the congruence of myself with the life process that exists outside of the fences of our society. Because in Nature I am stripped of blackboards, books, professors, television, newspapers, and hi-fi, a new voice, slowly but surely, has been able to establish itself and has asked me to respond to the callings of the land.

I can well remember touching my arm while on a broiling hike along the bottom of the Grand Canyon and feeling that my skin was cold from evaporating perspiration. I took a precautionary salt pill and thought, "How does my body know when to cool itself? How does it know when to stop, when it has reached the best temperature for optimally maintaining life?" I also remember the spectacular thunder shower late that afternoon and the cooling effect of the evaporating rain. "Why does the planet never get too hot or too cold for life here to exist?" I wondered, "How does it, like my body, know its temperature limits and how to maintain them? Does the same awareness and life process exist in me as in the planet?"

I watched the muddy Colorado flow (as it had for eons) carrying salts and sediments into the sea; I remembered "*Yet the sea is not full*"— ECCLESIASTES 1:7. "Under such circumstances, how does the sea maintain its salinity?" I asked myself. "How does it not become too salty? Is it the same process by which my body *knows* how to regulate the salt pill that I just swallowed? Why doesn't my blood become too salty from eating potato chips? What life force or consciousness keeps myself and the seas from becoming too saline? Do the seas rid themselves of excess salt in salt domes and deposits, some of which we find here in the Grand Canyon depths? How does my body know how much salt to excrete in order to maintain the life-giving equilibrium of my blood plasma and lymph?"

The Smoky Mountain forest floor and Iowa prairie soil seem to say, "The life cycle of people and the planet is identical." The rocks and sea beds that lie in mountain strata have found ways to explain to me how the planet recycles its soils that were once eroded from the uplifted continents.

The tangy scent of tidal flats and the south wind on the hillside insinuate that my body and lungs know how much oxygen to inhale or carbon dioxide to exhale in the same way that the planet knows how to maintain the proper balance of these gases in the atmosphere. The experience hints to me that the planet and my body maintain the proper balance of gases in order that life may continue undisturbed by either an overabundance or a scarcity of essential gases. Life is conscious of its needs and self-preservation.

An inquisitive part of my makeup asks a yellow birch on the tundra's edge, "Why are you living here?" The same part of me composes the tree's response, "Because I feel like it, it feels right." I mention this transaction at a seminar, and I am flooded with protests from students with ecology and biology backgrounds. "No," they chorus and then defend their positions.

"The facts show that the tree has genetic qualities which enable it to adapt to the harsh, icy, windy, wet climate at this altitude. Root competition, ability for its seedlings to grow in the shade of the mother tree, soil content, Ph factors, metabolism, and its physiology all play a deciding role."

To my students' surprise, I agree wholeheartedly with them, but I break down each of these facets into tension-producing or tension-releasing factors that the tree undoubtedly experiences on some level. As we talk, a cold breeze comes up. Feeling the tension of its wind-chill factor, we move our discussion to the protected leeward side of an immense glacial boulder nearby. The chilliness reduced, we continue our discussion, but our actions have spoken for themselves. I argue that our adaptation to the wind reduced tension. It felt good. We were more comfortable behind the rock, so we remained there. How do we know that on some level of consciousness the birch seed or seedling doesn't experience the same thing? Where's the solid, objective, scientifically controlled evidence which proves that mechanical physio-chemical reactions alone determine the tree's niche? Perhaps it is through them that the tree enjoys a less tense niche, just as we did. Perhaps the tree has some form of consciousness, and we have more in common with the tree than our upbringing lets us admit.

The Everglades and the Hudson River convey the message that I replace my waste fluid loss no differently than does the planet. The snow clouds over the Colorado Rockies, the gentle mists of Vermont, and the rains of the Olympic Peninsula show me that the renewing process that produces clean water is one and the same for myself and the planet.

A silent bog in Maine tends to ask me, "Isn't my peat moss buried deep in the bog away from decay and air for a reason?" The raw, exposed coal fields of Appalachia and Wyoming make the same query about the fate of the ancient forests that they once were. An ecologist provides the implications of the rising CO_2 content in the atmosphere due to industrial combustion, and acid rain from release of sulphur that was formerly trapped and contained in coal and peat deposits deep within the Earth. And although the specific words of each aspect of the planet say something different, they all sing the same song, whose music is the mutual self-preservation of the lives of people and the planet. For most Americans, it is a song whose beauty and message have been subdivided to the point where it is no longer recognizable nor sung as part of everyday life.

With much encouragement from each other, the expedition members become expressive of their thoughts and feelings about our intriguing visits with the Amish, a seashore, the Cajuns, the Sonoran desert, the Papagos, and many others. Our own widely varied backgrounds place each get-together in a shifting perspective. As we share our perceptions, we learn from each other's upbringing. The same geological wilderness or Morris dancer is experienced differently by a midwestern Protestant male student than they are by an Alaskan Catholic female or a Buddhist Floridian or a Jewish Vermonter. Much time is spent getting to understand each other's viewpoint in order to arrive at decisions based upon group consensus.

In contrast, it is serenely quiet as when we try to sense a northern woodland by being still for a half hour. The song of a thrush is carried by the sighing wind. The trees murmur, a cricket whistles, the brook chants its special call. I am reminded that music is found in every culture, that even with all its diversity, the expedition community is as one when it sings the songs of those who once lived here during another era. I have heard the silence of the mountains and recognize their deep stillness to be the harmony of the planet. I recognize that like the beauty of life, music is a gift that I share with the planet. We share the gifts of life because we are alive.

Although the ramifications of the planet being alive are controversial and staggering, the discovery is not new. Indications that the planet is alive have been around since the time that people first became conscious of the environment. Historically, people have universally conceptualized the Earth to be a womb. A womb is not dead.

Don't we feel comfortable with the symbols "Mother Earth," "Mother Nature?" Do we conceptualize a dead mother with these words?

Is God's spirit dead? If *the spirit of God moved upon the face of the waters* (GENESIS), are the waters dead?*

Throughout Genesis, life thrives on the creation of the Earth. There is no distinction in Genesis that declares the soil, mountains, and rocks dead. In the beginning, the planet was divided into inanimate and animate life. The only distinction was that some of the living things moved.

The Garden of Eden contained the essence of the planet. God put people in to *dress it and keep it.* Was it dead? Is it now? It has lost much of its ability to support life; the whole Mid-East, where once existed the Garden of Eden, has been exploited to a state of desert aridity. Yet, even in its present condition, it is not dead; it is a different form of life.

Let us hypothesize that the planet Earth is alive. The planet's ability to maintain life and act alive suggests that the planet *is* alive. Rocks and mountains, sand, clouds, wind, and rain all are alive. Nothing is dead, unless perhaps it is extinct. Even a doornail is alive—it is a manifestation of atoms and molecules in perpetual motion. Death as we know it is an illusion, if the Earth is alive; for if the planet lives, what can be dead? Death is actually nothing but the recycling of some of the energy of life.

§

---

*Many of the bible quotes in this book have originally been obtained by word of mouth from people we've visited.

Life cannot be defined. We can describe some properties of life common to life forms, but at this late date, we still don't know what *life* is. We often think and act as if *life* means *human life.* It's not life's complexity that makes it or a living planet undefineable. Life contains emotions, but our generalized stoic symbols make life seem inert, using dead symbols to define a living planet places the earth in a coffin within our mental images. We often fail to recognize that symbols like science, technology, progress, and culture have no substance, will, feeling, activity, or impact of their own. We may believe that planet earth is dead because "science says so" or "that's our culture." But culture and science are nonexistent; they are fabricated, security-seeking figments of our imagination, euphoric generalizations that along with our disregard of our Planet Blue feelings encourage the planet's imagined death.

Only after I had written an outline for this book of what Nature had taught me, did I skim through the literature. I was delighted to read that through the ages, some people have been able to see through the bias and influence of western culture. In living with Nature or studying Nature and the Earth, they too have been able to hear the voice of the living planet speaking in their blueness. Their thoughts were not part of my formal education even though I was a Life Science major. Some of them were scientists. I wonder what Ms. Gould does with their messages, which tend to confirm my observations....

Empedocles: *There is no birth in mortal things and no end in ruinous death. There is only mingling and interchange of parts, and this we call* Nature.

Black Elk, an American Indian: *Is not the sky a father, and the Earth a mother, and are not all living things with feet or roots their children? This pipe represents the Earth from whence we came, at whose breast we suck as babies all our lives, along with all animals and birds and trees and grasses.*

David Laing: *The Earth is not simply a warehouse of unrelated miscellaneous parts. It is rather, a functionally integrated system—as much an organic being as you and I—that exists not for man's benefit but for it's own.*

Aldo Leopold: *The land is an organism.*

John Lobell: *The Earth is a living organism capable of the self-regulation necessary to maintain the temperature range and complex chemical balances that support life.*

J. Donald Hughes: *Pythagoreans held the cosmos to be spherical, animate, ensouled, and intelligent.*

Rolling Thunder: *The Great Spirit is the life that is in all things—all creatures and plants and even rocks and the minerals. All things—and I mean all things—have their own will and their own way and their own purpose.*

Noel McInnes: *Earth is an organic spaceship. It lives.*

Wendell Berry: *Our bodies are not distinct from the Earth, sun, moon, and other heavenly bodies.*

Elizabeth Drew: *Sown in space, like one among a handful of seeds in a garden, the Earth exists.*

Harriet C. Childs: *Oh beautiful Earth, alive, aglow.*

Scholastic, Inc.:*Living Planet,* a book recommended by the National Educational Association and published by Johnson's Wax.

J. E. Lovelock: *We may find ourselves and all other living things to be parts and partners of a vast being who in her entirety has the power to maintain our planet as a fit and comfortable habitat for life.*

William Shakespeare: *The Earth is a thief that feeds and breeds.*

Lame Deer: *You have despoiled the Earth, the rocks, the minerals, all of which you call* dead, *but which are very much alive.*

Eugene Kolisco: *It is a peculiar fact that all the great astronomers of the 15th and 16th centuries were deeply convinced that the whole universe was a huge living being. Science is on its way to discover that all minerals originate from living things. Life never came into being on Earth, only dead matter has, out of the original life process.* (As fingernails and hair are produced by living people.— MJC)

Job 12:8: *Or speak to the Earth, and it shall teach thee...*

M.C. Richards: *The world is alive throughout. Even the processes of death are connected in the ongoing movement of human being and universe.*

Hermes Trismegistus: *The earth is a living creature, endowed with a body which men can see and an intelligence which men cannot.*

Gary Zukav: *We would like to think that we are different from the stones because we are living and they are not, but there is no way we can prove our positions. We cannot establish clearly that we are different from inorganic substances. Inorganic substances can make decisions, react to stimuli, and process information. To dance with God, the creator of all things, is to dance with ourselves. What we experience is not external reality, but our interaction with it.*

§

Nature is the epitome of life relationships; she is a fluctuating process that sustains equilibrium amongst her constituents—which include the living Earth. The varying T-R static and kinetic energies in the universe, the biosphere, animals, plants, minerals, and people all are variations on one theme: Nature. Each is alive and goes about the tension-release of living at different levels and metabolic rates, each according to its surroundings, history, and memory.

Susan Gould echoed Buckminster Fuller when she told me that the Earth was *like a spaceship.* That analogy arose from our unholistic emotionally estranged, encouraged, scientific and technological dependencies. A spaceship is not an organism. It is difficult to develop a feelingful interrelationship with a spaceship. One can generate more compassion for a salted slug.

Was the creation of the Earth merely the creation of a machine?

A more appropriate symbolization of the planet is that Earth is an immense, subtly pulsating, gentle, living cell, a gigantic Euglena; a total being that, with sunlight, is able to maintain life within itself. We are an organelle of that cell, one of its many systems. Its consciousness is the source of consciousness for the individual and collective life forms of the Earth. Nature has convinced me that life is conscious of itself. My own and every other organism's personal desire for self-preservation is part of the living planet's conscious desire to preserve itself. I have experienced myself being part of the planet's means of becoming aware of itself and its processes.

As a feelingful organ of consciousness in the vast living being that we call Mother Nature or Earth, human consciousness is both our own, and part of the planet's means of survival. It has become apparent to me, and to many others that whatever we do that damages our lives, invariably damages the planet, and *vice versa.*

I have learned from Nature that I am a natural being, that people are the personification of Nature, and on a subconscious level, that we have a relationship with and emotional tie to Nature which is as close and as meaningful as our relationships with family and friends. We are to Nature as a fish is to water.

The relationship is real, because the planet is alive and in many ways is us; the planet relates to us. The satisfaction we gain from eating food, drinking water, excreting wastes, appreciating art, making music, and having friendships is derived from Earth's positive contributions to its relationship with us. I observe pollution, cancer, shortages, and pain to be some of Nature's negative reactions to our unhealthy contributions to the relationship.

Many of us have difficulty seeing ourselves in this relationship because we are the eye that views it. An eye can see an environment, but it can't see itself.

I can no longer ignore that my ability to experience feelings has not just arisen out of nowhere. Feelings are an integral part of life on Earth. Protoplasm, the basis of cellular life, displays irritability. Feelings are a means by which cellular life experiences Nature. Feelings are consciousness of the build-up and release of tension, and express needs, desires, and motivations.

Because in our society feelings are subsidiary to the facts and figures of science, economics, and technology, I will continually remind the reader that each of us is feelingly colored Planet Blue, and that every element of Nature is within the newborn baby and communicates with the infant from within. The child knows Nature intimately, because it is Nature; Nature is all that it feels.

The pulse of Nature is the same as the life-giving adaptable heartbeat of an infant. The rivers, lakes, and seas of the planet establish contact with the newborn through the child's feelings of thirst and moisture. Through its feelings of thirst, the child knows of the existence of rivers, even though it's never been near one. The presence of trees, flowers, soil, wildlife, and human life is known to the child by its cravings of hunger and loneliness. The recycling needs of the planet communicate themselves as a need to excrete. The infant's feelings recognize the existence of weather, seasons, landforms, and gravity; they appear as desires for shelter and support. Our felt need for an ecological niche tells us that we have Nature's support and approval. The absence of a niche is felt as anxiety or anger. Messages from sunlight and climate are heard in temperament, exhilaration, and temperature changes. Nature's love for us and the reproduction of our species is experienced in our romantic and sexual feelings. We experience the pulse of Nature as self-preservation awareness, the process of tension and tension-release that is needed to maintain life and a living equilibrium with the living planet. Perhaps we know that God is alive because *By His work the Master is known.*

Nature is within the emotionality and unconscious memories of each of us, but we are acculturated by learning that our planet feelings (Nature) are bad and that distance from Nature is good.

Lame Deer:

> We live off nature, my wife and I; we hardly need anything. We will somehow live. The Great Spirit made the flowers, the streams, the pines, the cedars—takes care of them. He lets a breeze go through there, makes them breathe it, waters them, makes them grow. Even the one that is down in the crags, in the rocks. He tends that, too. He takes care of me, waters me, feeds me, makes me live with the plants and the animals as one of them. This is how I wish to remain, an Indian, all the days of my life, away from everything, to live like the ancient ones. On the highway you sometimes see a full-blood Indian thumbing a ride. I never do that. When I walk the road, I expect to walk the whole way. That is deep down in me, a kind of pride.

Like other forms of life, humanity carries the gift of consciousness and the anatomy to relate to it. It can be argued that evolution is not only a matter of the fittest surviving, but also an evolution of greater perception and consciousness of Nature. Whereas most organisms seem to relate directly to the forces of Nature, people instead become highly conscious of these forces and relate to their consciousness, which becomes part of their culture.

Slowly, as I have lived outside continuously, the planet has taught me that it is alive and how to sensibly and happily adapt my life to its life. I have discovered that the concept of a living Earth ties much together that has been in conflict for centuries, for it makes people, plants, animals, bacteria, and the solar system all variants of the same elements and processes, all facets of the same wholeness. They all are individual patterns of the tension-building, tension-releasing process of life. They're all but of one clay.

Susan Gould told me that there was no proof that the planet is alive, and from a purely scientific standpoint, this may be true. But a scientific outlook has yet to prove itself to be more universally worthwhile than other more subjective outlooks. True, western culture has brought us advanced technologies, but along with them have come immense social problems that seem to escape the problem-solving ability of the scientific method. It could be argued that the scientific method is not congruent with the way the planet operates because it is not holistic. It ignores the T-R emotionality of our life system. It is similar to claiming that a human statue that has no genitals is an accurate portrayal of a human being. Scientists have not been able to prove conclusively that the Earth is a living being. In order to prove that the Earth is alive, they need objective evidence—and there's the rub. The scientific process is incomplete in such an investigation, because in it the feelings of the scientist are taboo.

I have experienced that by respecting the Earth's and my own self-preservation feelings, we both thrive. That, it would seem, is evidence enough for a living planet, even though it is only an observation. It is also a key to offsetting the impact of our culture upon Nature. I often wonder what images or feelings went through the minds and hearts of the scientists and the U.S. Forest Service personnel who, in 1966, cut down an ancient Bristlecone Pine tree on Wheeler Peak in Nevada. Those with whom we spoke concerning the incident said that a researcher was attempting to tree ring date the tree and accidentally broke the bore drill in the process. In order to retrieve the $50.00 drill, the tree was cut. It proved to be close to 4800 years old. At that time it was the oldest known living organism on the planet, except of course, now it is "dead".

I remember being struck by the tree's plight, when we payed our respects to its stump as it lay in its alpine forest home. Somehow I identified with the tree, and I knew that at one time I could have been the scientist who killed it in the name of research or economics. I wondered where the source of my remorse lay, and why it did not come to play in the researchers so as to avert this tragedy. What was the ultimate origin of my concern about this incident, this tree, or its ancient neighbors? Wasn't my concern part of the life process? Wasn't it the planet itself trying to protect itself and, in turn, myself? Wasn't Nature the origin of my feelings? My life?

If each individual's feelings for self-preservation can be demonstrated to be part of the planet's survival consciousness and ability to preserve life, then the fact that scientists have these feelings would help them prove that the planet has survival consciousness and therefore must be alive. But science has little use for the validity of feelings; feelings are *subjective* and a contaminant to the purity of scientific investigation. By denying the existence and validity of their own feelings, scientists deny themselves some of evidence of the planet's life.

A corollary: If scientists were to work within the consciousness that the Earth is alive, they would be forced to include and integrate their self-

preservation and moral feelings about life into their pursuits.
It would become immoral to ignore their personal effects on the living
planet and other organisms including plants.

A scientist's moral feelings might curtail a generation of the
technologies and experiments that unnecessarily injure and kill the life of
the planet: the heartless vivisection of animals, the destruction of habitats,
the bombs, the poisons, the arsenals of weapons, the polluted land, water
and air.

A scientist whose consciousness does not accept the planet as alive
often is the scientist who participates in the planet's destruction. The
rejection by individuals of their own feelings of self-preservation can lead
to the partial demise of our planetary life-support system.

Like all of our other life processes, our feelings and consciousness
originate in the planet and are part of the planet. Our destructive feelings
about the planet or about ourselves lead to the planet's destruction as we
utilize *natural resources* to generate euphoria. Our thoughts and feelings
seem to serve the same function for the Earth that an individual's
conscience serves a living human being. We can consciously and
simultaneously effect our own survival and the survival of the planet. This
suggests that the planet is alive, because a dead planet, by definition, could
not have a living conscience.

Scientists and academicians worth the name will in time discover the
scientific value of their feelings and begin to experience that the Earth is
alive. Even now, some scientists are working with global environmental
problems that indicate that the planet lives. Although these scientists
continue to maintain their exclusivity (their scientific terminology calls the
living Earth *Gaia*), the existence of her life remains apparent.

The Expedition has visited scientists who have lent scientific support
to my own lessons from Nature. They have pointed out to us that Mother
Earth has for eons regulated the life-supporting salinity and consistency of
the sea, the life-giving proportions and contents of the atmosphere, and the
life-preserving temperature and climate of the globe. This she has done by
organically changing the weather, the clouds, and the color, shape, relative
size, and location of the continents and the ocean. On a global scale, she
has exhibited adaptive self-preservation behavior that is identical to that of
any living organism. She seems conscious of the parameters needed to
sustain life today.

In the future, scientists, academicians, and politicians will recognize
the personal congruency of their consciousness with the pro-life consciousness of the living Earth.

By using as a measuring stick the relative ability of an ecosystem to
support life, our civilization may someday scientifically discover that the

euphoria of living is on some level enjoyed by every entity that exists upon Earth, from the globe itself to its sub-atomic particles.

As we become conscious and accepting of the universality and importance of feelings in the life process, we will in time fully experience the joy of being alive and of maintaining the wholeness of life. I hope this will happen to a majority of us before we destroy humanity.

We must not wait for science or academia to tell us what our thoughts and feelings already tell us to be true about our lives, because most institutions are prejudiced against feelings. Our personal experiences have value and validity; our natural self-preservation feelings are unbiased as far as survival is concerned.

I have discovered that if I treat the planet with the care I would accord an animate living being, the planet responds positively, as if it is alive. When I treat it as if it is dead, it looks and acts dead. I would not pave over a living person or animal; that would kill it. Scientific or not, our attitude toward and treatment of the Earth is the difference in *aliveness* between the paved parking lot of a supermarket and the stunning oceans and green fields of America.

There is a great deal of history, logic and evidence to support the concept that the planet is alive and that feelings are biologically a vital part of the whole T-R life process. In order to survive, we must learn to think globally. We must add an emotional holistic lens to our cultural glasses in order to correct the warp that has produced our environmental problems. Even newspapers are willing to print editorials to this effect. (see appendix D).

Summary and Conclusion

The argument that we can't be prejudiced against a non-living entity is countered by suggesting that the planet is a conscious living organism because it acts like one. Over periods of geologic time, the conditions that define life—metabolism, reproduction, and adaptation—appear to function in the planet. Empirical and apriori reasoning suggest that if Nature is alive, the Earth is also alive. People and the planet exhibit and share the same life processes and systems including tension buildup and release (T-R), reactivity to feedback (homeostasis) and awareness on some level of T-R feelings and the desire to survive (consciousness).

Historically, the concept that the planet is a living organism has existed since the origins of culture. Wholeness, feelings and consciousness appear to be properties of life. Although some modern scientists have evidence that suggests the Earth is alive, their findings aren't conclusive because the scientific method usually ignores feelings; it subdivides and manipulates the planet and the scientist. Unholistic investigations tend to produce inaccurate conclusions about the existence of some form of life in inanimate objects.

Chapter Seven

CURSE NOT THE EARTH

Why does a natural area appear so foreign that we can't relate to it, to life, or to each other?

Blam! The sharp report of my .22 rifle was an exclamation point of sound as the bullet struck the woodchuck in the head. Through the 'scope I could see the bloody body heave convulsively as the woodchuck struggled and died. Arnie and Roger congratulated me. "Great shot!" they chorused excitedly.

A slight chill of guilt tinged my excitement, but I pushed back my feelings of remorse. "That's part of the game," I told myself firmly.

My army buddies and I were whiling away a Saturday afternoon shooting woodchucks in a pasture. The farmer who owned the land had urged us to shoot them—"They're pesky varmints," he'd said—but we weren't hunting to be of service to the farmer. We were hunting for the fun of it.

I was twenty-five years old, and fresh out of the army. To me woodchucks weren't a matter of ethics or morals. They were life, but they weren't too important. Shooting them was a way to enjoy myself, a way to try out my hard-earned basic training skills. I had learned to symbolize hunting to be a form of recreation, but my symbols were incomplete. They did not recognize the wisdom of the ages: *For that which befalleth the sons of men befalleth beasts: as the one dieth so dieth the other.* (ECCLESIASTES 3:19)

If there's anything that scientists agree upon other than that humans are the *highest* forms of life, it's that our self-preservation uniqueness in the animal world lies in our ability to symbolize.

We can see a tree in our backyard or in an illustration. Without understanding its ecological role, we can label it with the word *tree,* and thereby stick its image into our minds, where it flourishes not only as an image and word for *that* tree but for trees in general.

We relate to and through language and symbols of the life process. We relate to woodchucks and huge portions of the universe through the incomplete generalizations of symbols, and the present chaotic state of the world reflects this. It is as if we have invented a new kind of vegetable that is destroying our garden, because we have not invented a tool that can control its growth.

Right now, I'm nowhere near my cabin in Maine, yet I can see it in my mind's eye. Most people are very familiar with images of their house or car because they've spent a lot of time in them. We spend our whole lives in our own skins, yet our familiarity with ourselves and how we operate is negligible in comparison to what we know about the successful operation of a motor vehicle or house. We are missing accurate symbols and images that conceptualize our interest and need for self-preservation, as living organisms on a living planet. To discover how to relate to a wetland, to another culture, to our mothers-in-law, or to ourselves, we have to haul in a team of experts: ecologists, anthropologists, religious leaders, psychiatrists, philosophers, doctors, biologists, economists, and the like. We have learned to rely on their trained ability to symbolize to augment our own.

How did we lose contact with the normal, feelingful process of symbolizing to the extent that, in order for people to determine how to understand and relate to life effectively, we have to rely upon a group of experts who rarely agree with each other? When did the individual lose control?

Lame Deer:

> We Sioux spend a lot of time thinking about every-day things, which in our minds are mixed up with the spiritual. We see in the world around us many symbols that teach us the meaning of life. We have a saying that the white man sees so little, he must see with only one eye. We see a lot that you no longer notice. You could notice if you wanted to, but you are usually too busy. We Indians live in a world of symbols and images where the spiritual and commonplace are one. To you symbols are just words, spoken or written in a book. To us they are part of nature, part of ourselves— the earth, the sun, the wind and the rain, stones, trees, animals, even little insects like ants and grasshoppers. We try to understand them not with the head but with the heart, and we need no more than a hint to give us the meaning.
>
> What to you seems commonplace, to us appears wondrous through symbolism. This is funny, because we don't even have a word for symbolism, yet we are all wrapped up in it. You have the word and that is all.

Our inability to symbolize life accurately leads us to three problems that combine to overwhelm us and prevent us from having a healthier environment and society:

1. An unexplained, pervasive desire to *develop,* exploit, and destroy natural systems.
2. A distancing of people from Nature that prevents us from experiencing the impact of our daily lives upon the ecosystem and each other.
3. An inability to comprehend the millions of interlocking interrelationships that make up the complexity of the biosphere; a lack of emotional awareness that makes experts, no less than the average person, unable to fully understand and relate to Nature or self-preservation.

Why are we in this predicament? If the planet is being attacked by an unsolvable problem, isn't it more effective to search for the heart of the conflict than time after time to be enervated by fighting its daily effects? Isn't it vital that we recognize our problem's origin, if we are to resolve it effectively?

I have asked many people why they think we have increasing environmental problems. Over the years no practical answer has appeared while the problems in many cases have worsened. Why can't we discover their source? Because these problems emanate from a complex to which we are normally blind. I suggest that as unlikely as it may seem, the correct symbolization of the source is *prejudice against Nature.* Prejudice is a form of blindness.

Prejudice is not new to Americans. We have had a long history of being prejudiced against women and various minorities. I am convinced that this prejudice is connected to our relationship with Nature.

Ignorant of our prejudice against Nature and Nature's children, we exploited and annihilated the American Indians. We called it missionary work and economics. We *gained* a continent and *lost* ethics, wholeness, and pride.

The result of our barbaric relationship with the Indians has been a gnawing relentless conflict that is fueling many Native American fires today. There are those who say we never understood or appreciated the complex culture and equilibrium of the Indians. But unconsciously we understood it all too well. To us, it was wild, it was natural, and that was sufficient.

It is difficult to recognize that *What shall we do about the weeds?* is a prejudicial statement that is the same as *What shall we do about the Redskins?*—but it is.

We still exploit Blacks of Africa. Originally, we tore them out of their farms and communities. We considered them as natural resources and as such integrated them into our economy. They were useful as laborers so we used them.

The Blacks, like the Indians, were a people living harmoniously with Nature and were accordingly exploited. We rationalized our actions as legalized slavery, economic propriety and religious righteousness. When we forced the Blacks and the Indians out of harmony with the natural world, we saw ourselves on a mission of progress, not of prejudice. Certainly not of prejudice against Nature.

That Americans are unaware that they are prejudiced against Nature is to be expected. Prejudice is often an unconscious emotional phenomenon; aspects of it are invisible, and therefore not experienced or symbolized.

When it comes to Nature, most of us are as confused as the man who remarked about racial prejudice, "There are two things I just can't stand: people that discriminate and a damned nigger." We are prejudiced against being prejudiced; it is culturally taboo. We will reject prejudice, once it is discovered, but we are subconsciously unwilling to discern it, for to admit to being prejudiced is to admit to being *bad*.

Prejudice of any kind is a conditioned conceptual scheme. A person who harbors a prejudice appears to be unable to absorb additional symbols of perception from certain areas outside of his or her ingrained conceptual scheme. It would have been very difficult to convince me that I was wrong to shoot woodchucks.

To challenge any single aspect of a prejudicial configuration is actually to challenge the whole configuration of prejudice. The whole is glued together with culturally conditioned and resolved anxieties, trusts, and comforts, as well as the seemingly logical history of the configuration's development. Confrontation dislodges this emotional adhesive; we suddenly feel attacked, anxious, and defensive. That's how I would have felt in 1955, if I had been confronted in the pasture with the insinuation that by shooting woodchucks for fun I was prejudiced against Nature. (See Chapter 4, References 21 & 43.)

In order to prevent the emotional discomforts of confrontation, prejudiced people attempt to avoid self-exposure. Their escape from confrontation is often accomplished by avoiding issues, by associating with others of similar persuasion or by insisting upon the use of words and symbols that are not associated with the prejudicial configurations.

We experience no emotionality or conscious prejudice against Nature in the statement, "Excretion is a part of Nature." But the word *excretion* hides from us our uncomfortable cultural prejudice against Nature, against the natural process of excretion. Excretion is elimination, which is fecal material, which is feces, which is turds, which is crap, which is shit.

The latter part of the preceding sentence tends to be emotionally charged. The words are usually taboo, simply because *shit* is an *uncivilized* natural product, and a symbol which disruptively shakes out some of the emotional glue that binds our prejudice against shit, natural processes, and Nature. The resulting dislodged anxiety, combined with taboo reactions from others, soon conditions the cultured person to avoid using emotionally charged terms in *cultured* circles. We learn that Nature is wrong and culture is right. We do not experience this phenomenon with the word *desk*, because we are not prejudiced against office work.

It might be commented that the preceding example is invalid, because shit is a slang term—but slang is the language of *street people*, who are *uncultured*, i.e. less *civilized*, i.e. more barbaric, i.e. feral, i.e. natural...

Some of our language symbols' prejudice against Nature is less detectable and therefore more destructive than the previous example. For instance, consider the words "self-preservation" or "survival instincts." These are symbols that we commonly use to convey our desire to retain life, to experience the planet blue euphoric feelings of life that we first learned in the harmonious environment of the womb. Another way of symbolizing these same powerful, joyful, desire-to-live feelings is "love of life" or "joy" or "reverence for life." Our society is prejudiced against the use of these latter symbols because they get too close to Nature, too close to the euphoric feeling of *being harmoniously alive,* the essential feeling of life itself. People who speak of "love" or "reverence for life" are thought to be "hippies," "flakey," "counter-culture," or "fringe members of society." If "love" or "joy" or "reverence" were used in a textbook or scientific paper to describe self-preservation drives or instincts, that text would be suspect, as might be this one. Yet, it has been recognized that *Love worketh no ill to his neighbor; therefore love is the fulfilling of the law.* (ROMANS 13:10)

In order to bridge the anti-life gap between "self-preservation" and "love of life," I will use these phrases interchangeably in the text, so that our life-giving planet blue feelings won't be forgotten. They must be intensified. We must remember that our instincts for survival already have been weakened to the point that they can be overwhelmed by cigarette advertising, chemical use, toxic wastes, acid rain, and right or wrong, fighting for our country. We must remember that when we lose consciousness of our self-preservation feelings, we become suicidal.

§

Part II of my article *A Time To Be Born,* implies that our prejudice against Nature stems from the contamination of our daily rationality with an unconscious psychological bias.

We trust the innocence of the euphoria we obtain from our materialistic culture—but its innocence is not substantiated by a factual examination of its environmental effects. Our discrimination *for* material euphoria is prejudice *against* Nature. We attempt to avoid confrontations with the environmental facts, because they make us uncomfortable. Instead, we defensively enervate ourselves issue by issue, in time often losing *more* environmentally healthy ground than we gain, for no permanent solution has appeared. We avoid facing the facts by holding to our beliefs that our materialistic perceptions are *justified and thus not prejudicial.* As in the example of labeling shit *excretion,* which masks our prejudice against shit, justifying our prejudices enables us to lose sight of them for long periods until they are again confronted. I justified shooting woodchucks by telling myself I was participating in an *outdoor sport* and helping the farmer.

Lame Deer:

Let's sit down here, all of us, on the open prairie, where we can't see a highway or a fence. Let's have no blankets to sit on, but feel the ground with our bodies, the earth, the yielding shrubs. Let's have the grass for a mattress, experiencing its sharpness and its softness. Let us become like stones, plants, and trees. Let us be animals, think and feel like animals.

Listen to the air. You can hear it, feel it, smell it, taste it, **Woniya waken**—the holy air—which renews all by its breath. **Woniya, woniya waken**—spirit, life, breath, renewal—it means all that. **Woniya**—we sit together, don't touch, but something is there; we feel it between us, as a presence. A good way to start thinking about nature, talk about it. Rather talk to it, talk to the rivers, to the lakes, to the winds as to relatives.

You have made it hard for us to experience nature in the good way by being part of it. Even here we are conscious that somewhere out in those hills there are missile silos and radar stations. White men always pick the few unspoiled, beautiful, awesome spots for the sites of these abominations. You have raped and violated these lands, always saying, "Gimme, gimme, gimme," and never giving anything back. You have taken 200,000 acres of our Pine Ridge reservation and made them into a bombing range. This land is so beautiful and strange that now some of you want to make it into a national park. The only use you have made of this land since you took it from us was to blow it up.

The history of humanity unquestionably demonstrates that prejudice can exist when people relate to people.

However, history does not convey that hurtful prejudice also exists when people relate to Nature. Our prejudice against Nature has been covered up—justified—as growth, a conquering, progress, security, dominion, rise in the standard of living, prosperity, luxury, and increased economic gain. Teaching history in terms of these symbols helps to mask humanity's prejudicial relationship with Nature. Again: *when people believe that their perceptions and frustrations are justified, their actions and attitudes do not appear to be prejudicial.*

Historically, we have deprived Indians and other minorities of their life support opportunities by excluding them from jobs, housing, education, and land. These are acknowledged acts of prejudice against people. Similarly, to deprive the planet of its life support is an act of prejudice against Nature. From our ecosystem we arrogantly borrow food, air, water, and life—often returning poison or nothing immediately useful. We have not learned the scripture that states: *The wicked borroweth and returneth not again.*

Although we know intellectually that the biosphere is a living system with an integrity of its own, we treat other forms of life as natural resources and annihilate them to selfishly increase our *standard of living* or *quality of life.* Surely this is prejudice against Nature. Depriving plants and animals of their habitat or life without good reason is no different than killing people we dislike. It is a discriminatory statement that our luxuries are more worthwhile than they are—it is blatant prejudice.

Just as it is an act of prejudice to judge women or blacks unfit to use certain public facilities or to vote, it is similarly an act of prejudice to judge certain wildlife as unworthy to use the environment. It is prejudice against Nature to hold irrational suspicions and fears of soil, water, sunlight, weather, plants, and wildlife. It is prejudice against Nature to fortress ourselves indoors, away from them, as if they were *nasty* and unfavorable, or to bulldoze them into oblivion to make way for the building of an amusement park.

Our jaundiced eyes perceived Native Americans to be heathen savages; to eliminate them, we robbed them of their existence by robbing them of their land. It is prejudice against Nature to persecute people because they live closer to Nature.

Lame Deer:
Maybe it was a good thing that they would not let us

Indians keep that land. Think of what would have been missed: the motels with their neon signs, the pawn shops, the Rock Hunter's Paradise, the Horned Trophies Taxidermist Studio, the giftie shoppies, the Genuine Indian Crafts Center with its beadwork from Taiwan and Hong Kong, the Sitting Bull Cave— electrically lighted for your convenience, the Shrine of Democracy Souvenir Shop, the Fun House—the talk of your trip for years to come, the Bucket of Blood Saloon, the life-size dinosaur made of green concrete, the go-go gals and cat houses, the Reptile Gardens, where they don't feed the snakes because that would be too much trouble. When they die, you get yourself some new rattlers for free. Just think: if that land belonged to us, there would be nothing here, only trees, grass and some animals running free. All that **real estate** would be going to waste.

It is an act of prejudice against Nature to believe that God is more concerned about our technologically powerful society than He is about more natural cultures, or Nature. It is prejudice not to recognize *that a man hath no preeminence above a beast: for all is vanity.* (ECCLESIASTES 3:19)

Slavery, lynching, rape, and hostile lawmen were the legacy of Southern bigotry against blacks—which we recognize as extreme prejudice against people. It is echoed by the extreme prejudice against Nature evidenced in the legal poisoning, raping, and murder of Nature with pesticides, pollution, power dams, oil spills, acid rain, toxic wastes, radioactivity, and exploitation.

That *Mother Earth* can be unnecessarily *violated* or *raped* to lose her *virgin* forests is an application of prejudice against females transferred to Nature, and vice versa.

We are so prejudiced against Nature that we often need psychiatric assistance to discover and accept the inborn feelings of Nature within ourselves.

Our anti-Nature bigotry has driven us to conjure and believe in myths which fantasize that people were created by a power other than Nature—a power who created us in His own (therefore human) image.

§

These are only a few illustrations of prejudice against Nature. It is as pervasive as money. Once identified, it can easily be found in every inhabited corner of our nation. Prejudice against Nature can only be identified by confronting its justification.

Every *justifiable* act, thought, or feeling must be scrutinized, and measured against the pulsating equilibrium of Nature. Only in this way may we discover, label, and ultimately deal with our prejudice against Nature.

Our prejudice against Nature distorts our mental processes to the extent that we are virtually unable to find a congruency between the way the world works and the way we perceive it to work. There is overwhelming evidence that people have a relationship with Nature which is the basis of our survival. It is our prejudicial symbols, thoughts, and feelings about that relationship, which threaten our self-preservation.

We live on the edge of a great unknown. Self-preservation consciousness is a new frontier. It is a part of ourselves and the Earth for which we have no strong, positive symbols. The symbols that we do have now are prejudiced against Nature.

Our words are symbols of our relationship with the environment and with ourselves. For example, let's make the word *yes* mean *no* in the following situation:

I ask you if I can put poison in your food; (meaning no) you say yes, and you watch me do it. I ask you if you want to eat the poisoned food; (meaning no) again, you say yes, and I serve it to you, insisting that you eat it as you promised.

But do you actually eat it? No, you don't. Something else takes over: feelings of self-preservation. Words be damned, you don't knowledgeably eat poisoned food no matter what anybody says, no matter if your honor is at stake, no matter how hard I might argue that you keep your promise to me or your commitment to my cooking.

Self-preservation feelings and consciousness are natural prime factors in survival and in relating to Nature. But because they are natural, we are prejudiced against them and ignore them. How many college catalogues offer even one course in self-preservation? How much education, training, or reinforcement have we been given to open the valve of our self-preservation feelings? In comparison, how much exposure have we had to symbols which reinforce our culture—and often poison us?

Prejudicially, modern American civilization cultivates in us an insecurity in our innate Planet Blue ability to survive, and then instills in us the conviction that we must desire certain technologies, and take certain poisons, in order to survive in the culture. Our self-preservation feelings and consciousness have grown so weak that self-destructive behavior is commonplace and in some cases has achieved chic. Excessive amounts of technology. Alcohol. Drugs. Cigarettes. Immorality. Dangerous cars. On a larger scale, food laced with carcinogens. Pollution. Toxic wastes. Habitat destruction. Be cool!

§

The most subtle and devastating illustration of our prejudice against Nature is found in our academic and scientific theories, the words and numbers that we use to relate to Nature, to impose a controlling order of our culture upon the pulse of Nature.

Nature is rarely included in our symbols, even when our symbols purport to convey the *laws* of Nature. Has anyone ever *asked* Mother Nature if she actually has any laws? If Nature is a randomly organized series of relationships, she would not operate through laws but rather through respect or cooperative, sensitive, mutually beneficial tension and tension-release relationships.

The *pure* science of mathematics is a discriminatory juggling of Nature's tension-release process. It is ingrained into every school-age American child mandatorily. Mathematics is often a weapon that is used to subdivide Nature.

Nature has no knowledge or use of numbers; they are a figment of our cultural imagination. Numbers symbolize divisions of the whole. They are symbols that attack the natural integrity and value of Nature's wholeness. The use of numbers to abstract and quantify paves the way for exploitation of Nature in our search for technologic euphoria.

Mathematics leads us to notoriously inaccurate conclusions. Does $2 + 3 = 5$? No, it does not. It is incomplete. Look at the question again. Where do you see Nature, the wholeness of life, or people's self-preservation feelings? They are not there.

Two apples plus three apples equal five apples? No; in reality, two apples plus three apples *plus* the T-R wholeness of Nature that is necessary to produce apples *plus* the human inclination, energy, and individuals needed to conceive of gathering the apples to count them, then to count them; all that equals five apples. That's the full equation, because it includes the life of the planet and people.

To be fair to Nature, and to approach accuracy and validity, the wholeness of life energies of the planet must be represented in an equation. Mathematics often fails to recognize that even one apple is a statement of the eons of life processes and energies which have culminated to form it.

I symbolize the role of Nature in any equation by using the symbol *WL,* which stands for the T-R energy bonds, life experience, and evolution of *Whole Life,* since the beginning of time. WL is the Whole Life Factor. It is a symbolization of the concepts that I have presented in the previous chapters, an amalgamation of:

- the biologically vital significance of feelings;
- the highest probability of survival occurring when life systems are fully intact;
- the planet Earth being a whole living organism;
- tension-release relationships being the essence of Nature;
- consciousness being an awareness of the tension-release process in people and the planet;
- womb euphoria being psychologically addictive;
- the relative ability of the surrounding ecosystem to support life being a measuring device for any single life system;
- people being an organ of the organization of the Earth and subject to its tribulations;
- sensing the wholeness of one's blue-planet self being a sensing of the wholeness of the planet;
- the murdering of one's own life support system being the essence of madness;
- a person being the product of two mothers: a human mother and Mother Earth.

In order to avoid prejudice against Nature, the original equation of my example must read: 2 apples plus 3 apples = 5 apples *minus **WL***. If the planets' natural and cultural forces use up the energy of one apple, the real equation is: 2 apples plus 3 apples = 4 apples. For example, in a whole life perspective, an apple that I buy at the market in Maine may have been grown in a plowed-up Oregon wilderness that has been herbicided, insecticided, and fertilized. The apple may have been picked by exploited workers, unnecessarily cultivated, transported, packaged, and preserved with oil-powered machinery, then manipulatively advertised and marketed through several middle-merchants. The energy and questionable intents of these processes may account for 80% of the apple's cost. More important, each of these processes may have had an adverse effect on the whole life process of the planet, but this impact does not appear in the equation 2 apples + 3 apples = 5 apples. If *2 + 3 = 4* sits uncomfortably with your perception of mathematics, your prejudice against Nature might be coming to the surface and signalling in you the presence of an ingrained, prejudicial mathematical configuration of consciousness. The educational system which taught you mathematics may have taken the life efforts of Nature for granted. One has only to go into a forest and attempt to repeat the process of proving that *2 + 3 = 5*. Each rock, stick, plant, animal, or bit of soil that is manipulated by the proof, or by one's presence, must be part of the equation, if it is to be accurate. Each position change or removal in the forest changes the forest. Additional energy is needed for recovery to take place. That energy is part of the WL factor.

What happens when we work with equations that ignore the importance of the system of energy forces symbolized by the *Whole Life* factor? Making decisions based on the misinformation of incomplete equations has brought us pollution, habitat destruction, toxic wastes, energy shortages, and most of all the other social and environmental dilemmas that plague our culture today. These ills are a result of our having divided Nature's wholeness, broken her life bonds, and disrupted her life process, and therefore our own.

Mathematics and all other sciences which broadcast a similar discriminatory message—a message that ignores Nature—carry these problems in their wake.

Our schools do not teach the WL factor. They do not teach us that historically two plus two might equal three—or even two. To get back to basics, we must teach that two trees plus three trees does not equal only five trees, but equals also hundreds of branches, thousands of roots, and millions of interrelated organisms and life processes that maintain life, which we ignore in order to get good grades, a diploma, lumber, and rich.

Our system of numbers and equations must be made to account for the WL factor, if it is to be accurate and wholesome.

Geometry exhibits and instills the same prejudice against Nature. Most of us believe that the shortest distance between two points is a straight line. Why wouldn't we? The maximum has been drilled into us since kindergarten and before. But where is Nature or self-preservation in that determination? I can't find them. The facts are that the globe is circular, that Nature fluctuates, and that people are an integral part of the vibrating wholeness of the planet; there is little that is *straight* about the life process. A straight-line philosophy ignores all this, yet we are taught to believe that it is a basic truth. The real truth is that a straight line relates to Nature as the Nazis related to the Jews. We accept and impose straight lines on cattle with a red-hot two-bar branding iron, like stripes on a prisoner's clothing. Rain forests are chopped into straight broom handles when a crooked backyard alder can be used just as effectively. Railroads, skyscrapers, factories, houses, highways, and the professions that create them, straight-arm Nature and produce the circular problems that we now face. Vicious circles are caused by avoiding the tense feelings that have been ignited by our straight-line conquering of Nature.

Straight lines can be unhealthy. The WL factor tells us that there are very few straight lines in Nature. The shortest distance between two points usually is through a life system. Instead of making room for lives by going around them, our passion for speed and straight lines has prompted us to bulldoze and destroy those people and places caught in the path of *progress*.

§

It is extremely significant that modern physical and chemical *laws of Nature* cannot be verified unless one changes and divides Nature's wholeness. We have given our hurrahs to the physicist who discovered the *natural law* that in a vacuum, a feather falls as quickly as a pound of lead. In a vacuum, perhaps it will. But a vacuum is unnatural; it is a manipulation of the wholeness of Nature into a people-made vacuum environment. It may truly be said that the vacuum is in our mind. Only in this unwholesome, completely contrived setting can we enjoy the discovery of one of the alleged truths of Nature.

The real truth is that the WL factor of the great chain of life is missing in the many laws of modern physics and chemistry. If Nature is randomly fluctuating tension-release relationships, then there are no real *natural laws* or *standard conditions of temperature and pressure* upon which such laws may be based.

The real problem is that most scientists are taught to be mechanical and reductionist. They're taught to isolate, to control, and to conceptualize, but not also to apprehend the totality of things and see how they all are interconnected. The difficulty is that many scientists try to operate in a vacuum, just as in their experiments. Power dams, insecticides, toxic wastes, chemically-based food "products," *un*civil engineering, and Three Mile Island are among the consequences.

On a gut level, we can resent that our cultural bias has taught us such misinterpretations and misrepresentations of the natural life process that sustains us. But gut-level self-preservation reactions are generally held worthless in our civilization. Feelings are Nature. Feelings don't count for much in comparison to the *objective*, ultimate truths of physics, math, and science—especially when the latter are expressed as *cost-effectiveness, cost/benefit ratios,* and *engineering feasibility* "studies."

We are taught to be objective by teachers who have not been taught that the planet may be alive and the implications of that discovery. Many teachers do not know that, objectively, it is reasonable to be subjective when self-preservation is at stake, because survival is an emotional issue. We do not learn that the definition of objectivity includes dealing with subjective feelings. Instead, for example, in a recent three day teachers' conference, objectivity became a matter of examining facts and issues. I was objectively bombarded with over 184 mind-boggling facts within a four hour period, and they continued to spew forth for an additional two days. My congresswoman says that's what she experiences daily from lobbyists. Frankly, I don't completely trust even one fact, no less the objectivity of the person presenting it. Too often, I have been misled by the unholistic objectivity of scientific research and inaccuracy of facts in my own lifetime. To witness:

> *Fact:* In 1945, the scientific community certified the safety and effectiveness of the "fluoridation of water. This was a question that I got "right" on my 1945 Biology final.

> *Fact:* In 1982, parts of the scientific community believed that fluoridation might be a health hazard, and that people have served as guinea pigs in order for this fact to become available.

> *Fact:* Curtailing the use of sugar and bleached flour is a healthier solution to the problem of dental caries than is fluoridation.

> *Fact:* We would have a food shortage without bleached flour and sugar.

> *Fact:* In the name of objectivity, considering and evaluating the issues and facts of fluoridation produces hundreds of conflicting facts whose composite effect is to cause confusion and apathy.

It is also a fact that the evaluating process gives power to a small minority that are scientifically, academically, or commercially specialized to organize and lobby the facts that inadvertently manipulate people. This fact-oriented process is not objective, because prejudicially it usually does not include an individual's or the planet's self-preservation feelings with regard to fluoridation or most other subjects.

For some life-preserving reason, the planet does not fluoridate the water, or grow cane sugar in New England. It is within Nature's and our own self interest to find Nature's way to survive here in the Northeast without prejudicially altering the environment. We know that before we arrived, life existed here for immense amounts of time with neither sugar cane nor fluoridation.

All indications seem to say that life was stronger and more lasting when blended with the Indian's culture rather than with our bias toward excess technologies and distance from Nature. Isn't it more objective to trust the proven life system of the planet than the prejudicial schemes of our culture?

Our objectivity is in many ways reminiscent of the toad who desired to eat a centipede but could not catch her—at his approach, she would retreat up a tree. But she listened as he flattered her saying, "Oh wise, reasonable, beautiful centipede, how I desire your acquaintance." The centipede remained still as the toad's compliments beguiled her. Knowing he had her full attention, the toad said, "But explain to me, oh sensible, lovely centipede, explain to me how you decide which foot to move when you walk. Does foot #97 go forward when #38 goes backward? When #64 is still, do #82 and 15 go right or left? When both or either of the above occur, how do you objectively decide to more #23, #75, and #44 up or down?" The flattered centipede listened and became conscious of actions that before had been automatic, self-preservation instincts.

Slowly the toad began moving toward her while continuing his questioning. The centipede wanted quickly to retreat, but in order to maintain the toad's flattery, began to reason to herself with the toad's objectivity. "In order to escape the toad and survive, I must climb the tree. In order to do so, should I move #87 forward or #43 back? Oh dear, is it best to turn around by reversing #8 and #52 and #98 or back up with #29, #42, #71 and then move #63, #2—" CHOMP! GLURP! The centipede was toad food and was later excreted as a toad stool.

If there is a moral to this story, it is that in order to be fully objective in our society, it is reasonable to act off our feelings when self-preservation is in any way at stake. In America, due to our prejudice against Nature, *Nature is exploitable until proved necessary.* Self-preservation feelings must first become freed and encouraged to a level of intensity and creativity whereby they equalize the impact of the prejudices of our culture's "objective" facts and figures. Only at that point can objective examination of the issues begin in a democracy, because that is the point at ·which a vast majority of people are not apathetic. They are concerned about their own and their children's health and lives. Their numbers and the intensity of their participation becomes as powerful a factor to reckon with, as is any other aspect of an issue. We too often forget that most issues, especially environmental issues, are basically issues about long term survival, not only of economics or standard of living.

When self-preservation feelings are active, they can help offset the allegedly "objective" practice of placing monetary values on life, such as $50.00/duck, $140.00/deer, $120.00/tree which are then totalled and entered as calculations in the cost-benefit ratio for determining cost-effectiveness of a proposed power dam, nuclear power plant, or other technologically based project. This process is absurd. It is prejudiced against Nature, and does not recognize the Whole Life factor. It contains all the objectivity and reasonableness of hiring a deaf person to explain the values of listening to the radio.

Some people question the views of my wife and me simply because we have freely chosen to live out of doors, non-electrically, without indoor plumbing and central heating. We enjoy rural living, yet we are suspicious of an individual who is different from us, who is dirty or who eats with his/her hands or uses drugs.

We all are prejudiced. To believe that most Americans can be objective about their relationship with Nature is similar to believing that historically men have been, and are, impartial in their relationships with women. In order to be truly objective about our relationships with Nature, we must apply the WL factor. We must fully recognize that the T-R emotionality of our culturally stifled planet blue feelings *has never disappeared.* Instead, it has been subconsciously attached to each aspect of our accepted way of life, including our desire to be objective. If we are biased about a subject, can we really be objective about it?

Our prejudiced objectivity can be understood and rectified by considering the Whole Life factor and who or what we are at this moment.

Through extended contacts with Nature and people, my perceptions of Nature have changed from what they were 20 years ago. The expedition has helped me realize that, at birth and before, each of us was a miniature blue globe, a ball of life that replicated and is part of the life system of the planet. At birth, we were placed in the vise of our culture and society there to spend our first defenseless eighteen years of life. The vise has compressed blue-global us into cubes that are colored Cultural Red. We retain our red-cubic shape, even though as adults we have been released from the vise of childhood.

I recently saw a young boy being wheeled around in his mother's shopping cart. He cried after having his hand slapped for grabbing a box of crackers from the supermarket shelf. (Crackers that no doubt were placed at the height of the shopping cart in order that little boys could reach them and ask their mothers to buy them.) After a while, his crying stopped, but what happened to his hurt feelings? Where did they go? Did they just vanish?

As the previous chapters suggest, the little boy's attraction to the crackers was the expression of one or more natural functions with which we all are born. Through rejection, punishment, or social acceptance, we learn that these natural feelings in their natural state are wrong; there is a right and a wrong way to sneeze, eat, excrete, drink, breath, move, talk, have sex, relate, or get crackers.

Each of our natural feelings are hurt, to a greater or lesser extent, as they become socially trained. We become red-cubes by beating, bending, or cajoling our global, blue-planet feelings into acceptable patterns and attitudes. The hurt that this acculturating process generates does not disappear. Along with other hurts, it lies buried just below our red-cubic walls of consciousness and is felt when cookies again appear on the scene. It serves to inhibit cookie-grabbing.

Our acculturated hurt feelings remain hidden from consciousness behind our red-cubic walls until such time that the walls are ruptured by strained social relationships. Then hurt feelings squeeze out through the rupture, and we experience them as they come into consciousness. This is why we may feel uncomfortable when we are judged wrong or inadequate by others, or when we are criticized because we don't meet the red-cubic expectations of the boss, parents, teachers, policemen, acquaintances, or other authorities. We may even feel hurt when we don't meet our *own* cultural expectations.

In order to avoid the discomforts of experiencing our hurt or unhurt natural feelings, we spend much of our lives being "good," "obedient," or "competent." This strengthens our red-cubic walls so that they won't leak our hidden discomforts. Thus, even our red-cubic scientific "objective" attitudes and institutions are neither neutral nor objective. They are prejudiced against Nature because they unjustly fear and discriminate against our natural or hurt blue-global feelings.

To write a congressman, confront a polluter, change our lifestyle, or contact a neighbor about an environmental problem is a risk, because we know it could trigger negative reactions toward us. Critical reactions could weaken our red-cubic walls and let out some of the hurt that we carry. The formation of close relationships, by letting people honestly and openly know our feelings, assumes the same risk. Too often, we refrain from both. Instead of making red-cubic room for Nature, we turn to cigarettes, alcohol, drugs, and other escapes, because they tranquilize the hurt feelings that constantly seep through weak red-cubic walls, or they tranquilize the walls in order that natural feelings may be experienced safely.

85% of the population is concerned about America's poor relationship with Nature, because we know it to be the cause of our toxic, deteriorating environment. But most people are unable to act to improve this relationship. They're blocked by apathy, a red-cubic anxiety. Most of our days are spent reinforcing and strengthening our red-cubic walls, not transcending or changing them. Subconsciously, we know that *if we extend our hand, it may get slapped.*

We are not the free, just, equality-loving, pioneering, *objective* Americans that we have been told that we are. Instead, we are emotionally charged red cubes that have subconsciously learned that Nature is uncomfortable and harmful, but that cubic-red is euphoric. *We treat Nature as we were treated when we were Nature.* We grow up to become the vise. We compress the natural world, and children, into non-threatening or euphoric elements of our relationship-deficient, red-cubic lives. And until we apply the Whole Life factor, we suffer, along with the planet, the anguish and environmental consequences of our internal and external prejudice against Nature. Presently, we are unaware that, like the forests, our mentality has been prejudicially sprayed with red-cubic poisons in order to control and deaden the wilderness values that live within us. We can discover and neutralize these poisons by applying the Whole Life factor to them.

The democratic process allows us to ask the residents of a rich, all-white community to decide "objectively" whether to allow Blacks to live there, and thereby risk a decrease in property values. Is that democratic objectivity or unjust discrimination? The true test of our "objectivity" is to observe its end results, i.e. the long term state of the environment, our ability to survive. The results speak for themselves in all-white neighborhoods and on our deteriorating planet. Isn't our polluted air the foul breath of our "objective" way of life and of our present relationship with Nature?

Unless we are aware that we are "objectively" prejudiced against Nature, we will not attempt to deal with the problem by applying the WL factor. That would be unfortunate, because a great deal of headway has been and can be made in subduing the effects of many different prejudicial relationships. In order to objectively think globally, we must learn to incorporate thinking subjectively.

§

Even the most stupid individual must on some level be aware that Nature is not totally harmful and is the source of life for people. Only in rare instances is Mother Nature poisonous or lethal to the person who is knowledgeable of her ways. Life in the country, whether it be hunting, gathering, farming, working, or simply existing, has merit. Each year over thirty million Americans visit Nature in her last stongholds, the National Parks and other sanctuaries and refuges. Not only do people live through the experience, but they actually enjoy it! Isn't it interesting that when people have a vacation they often vacate America's more *civilized* areas and recreate (re-create themselves) in the country, close to Nature.

There are reasons why we are not able to retain the wholesome vacation values that we find in Nature, and thereby live closer to Nature's fluctuations all year round. One reason is that the strength of the values that we experience in Nature is masked and subdued by the symbols of our historic prejudice against Nature. Almost every negative, disdainful, or profane word in our vocabulary can be traced back to a definition approximating *too close to Nature,* or *not moving away from Nature.* Our negative symbolizations and attitudes are a seething mass of anti-Nature propaganda—and, like all effective propaganda, it is rarely recognized for what it is, unless we apply the WL factor. For example: *heathen* means *uncivilized.* It stems from *heath;* people who lived on heaths were thought to be uncivilized. *Pagan* stems from the Latin word for country dweller. *Savage* and *boor* mean the same as *barbarian;* barbarian means *a foreigner;* and a foreigner is *a person of the outdoors.* A *villain* is a *depraved, base-minded scoundrel.* The word *villain* comes from the Latin for a person who lives in a *villa*—a country house.

Many demeaning words, such as lout, uncouth, brute, illbred, and redneck, denote that the people so labeled are peasants; they live in the country. This really aggravates my planet blue feelings. Every year, for more than a decade, the institute lives for a time with peasants. They are the *plain,* or *team* people, the Amish and the Old Order horse-and-buggy Mennonites.

The Amish and the Mennonites live close to the land in small farming communities and are institutionally committed to following the traditions and ways of the European peasant class of the 17th century. They are Christian Americans who thrive in a manner that does magnificent justice to the word *civilized.*. They may have brought self-sufficiency, honesty, cooperation, and trust to their highest definition in America today. For three centuries, they have lived and enjoyed values that we call idealistic and yet cannot achieve ourselves on a community level—much less experience nationally.

One of the greatest strengths of the Plain People is their purposeful restriction of the use of technology. This institutionally-imposed curb decidedly limits their ability to even conceive of obtaining superfluous distance from Nature; it limits, too, the need to cope with the social and environmental problems resulting from such distance. They practice humility. They say, "That which is highly esteemed by men is an abomination in the eyes of God."

It is worth considering why western civilization has harrassed and mercilessly persecuted these people, who at this very moment practice peaceful, voluntary simplicity amidst the roars and wars of our materialistic society. I am convinced that they were and are shunned and persecuted not only because of their religious convictions, but also because they are peaceful and live close to the pulsating harmony of Nature. We demean them because, by remaining strong, their lifestyle is successfully, uncomfortably confronting our prejudice against Nature.They have told us that in their bad dreams or nightmares, they become one of *us*.

Today, some people from the business community demeaningly refer to the plain people as *honkies*. Although we have a few derogatory terms for each group of people against whom we discriminate, there is an exhausting number of derogatory symbols that defame people who have lived close to Nature. These defamations are not merely *a matter of semantics*. They are a matter of prejudice against Nature. Appropriately enough, I gathered information about these symbols from a book entitled *The American Heritage Dictionary*. Abetted by our unthinking use of the symbols listed between the dictionary's covers, prejudice against Nature has become our American heritage. And nowhere in its nineteen definitions of *culture* and *civilization* does the dictionary convey that these words connote a relationship to self-preservation or Nature.

Our culture's discriminatory anti-Nature symbols are hypnotically conditioned into our American mentality. They are imprinted on us during childhood, and they are continually reinforced because we are surrounded by them.

§ .

This morning, because the expedition bus wouldn't start due to a corroded battery terminal, it appeared that we would miss our ferry connection to Martha's Vineyard. We jump-started the bus by organizing the students and staff to push it. Those who were keeping diaries undoubtedly noted and described the unusual incident, yet I'm positive, because I checked, that many students had different ideas as to why we pushed the bus (dead battery, broken starter, lost keys, bad coil, corroded terminal). I'm sure many of the students noted in their journals just that we pushed the bus to start it, and will not include the reason why. Whose journal will future historians find? Which historian will produce the historically authenticated fact that Audubon Expedition buses were started by pushing them? Will the real reason for their being pushed ever be known or conveyed? And won't the in-vogue historian conjecture from his/her experience or frame of reference as to why the bus was pushed? Isn't it possible for that historian's idea to become historical fact, and for the future value of bus transportation to be based upon that historian's "history"?

The expedition only recently returned from a 3-5 million dollar archaeological restoration now in process, which is intended to further reaffirm the prowess of the whaling industry. We have visited a multi-million dollar whaling museum that stimulates the local economy and gives people pride in their whaling heritage, but fails to acknowledge the present endangered plight of many species of whales. The expedition has attempted to apply the Whole Life factor to history and to that museum. Our heroic attempts to have a whaling history museum convey the demise of whales has met with the same resistance and reasoning that was offered by Susan Gould toward this book. The museum's directors stubbornly insist that their institution is dedicated to the *history* of the whaling industry. As such, they state that they are in no way responsible for conveying that industry's contribution to the extermination of whales or of the museum's educationally negligent role in that respect. Yet the museum is given non-profit status and support on federal, state, and local levels, and is academically accepted as an outstanding, historically accurate institution. It is historically accurate, since it includes our blind prejudice against Nature.

By applying the Whole Life factor, we have found that history's bigotry in portraying whaling is no different than its defamation of Nature while conveying the "history" of agriculture, the conquering of the West (sic), the settling of America (sic), the industrial revolution, man's rise to civilization (sic), ad nauseum. We not only harbor the memory of hurtful deeds, thoughts, and feelings about Nature, but we reinforce them by applauding them as our heritage and history. With little regard to their past and present adverse effects, we educate ourselves and surround our minds with historic idols and ideologies that in many cases are fabrications. We allocate enormous sums of money, time, and energy to praise our conquering heros, politicians, and businessmen. A prime example of this is a State of New Hampshire Historical Commission road sign standing by a lovely stream. The sign reads:

> BAKER RIVER: Known to the Indians as Asquama-
> numauke, the nearby river was renamed for Lt. Thomas
> Baker (1682-1753) whose company of 34 scouts from
> Northampton, Mass., passed down this valley in 1712. A
> few miles south, his men destroyed a Pemigewasset
> Indian village. Massachusetts rewarded the expedition
> with a scalp bounty of 40 pounds and made Baker a
> captain.

The sign seems innocent enough. It is a typical presentation of "history," and history is important enough to be required for high school graduation. It is more difficult to perceive that the sign is an act of prejudice against Nature. If our living planet could speak, what might it say about it? Let's listen to the planet for a moment:

"I despise the Baker River sign. It is a hostile act against me. I'm not even sure that it is accurate. How do I know what Thomas Baker did 250 years ago? Are the journals that convey the information accurate? And

even if I did know, what difference does that make today? The sign conveys more than just an historic fact. The sign itself is history. It is a continuation of America's historic disregard and disrespect of the life process of the planet.
"The very presence of the sign is a prejudicial statement. It says that signs, history, and words are more important than the unadulterated beautiful scene of a living stream flowing through the mountains. The sign interrupts the scene in order to say that humans have placed inaccurate symbols on the stream. People have discriminated against me and made the stream become a monument for some army lieutenant named Baker. It is legalized graffiti. You value his past deeds over my present life. The deeds you value are acts of prejudice against me: the killing of life, and the extermination of Indians who lived close to me and revered me, who knew me as their Mother and the Great Spirit of life. To you they were heathens, because they demonstrated that your Judeo-Christian lifestyle was based upon living apart from me and was not a necessity for survival. I and the Indians were and still are a threat to you, and you, therefore, act accordingly—that's really why the sign is there. The fact that that sign states that this Baker fellow and his men were monetarily rewarded for their act, and that they evidently took scalps from the Indians they killed, does not speak well of your attitudes toward life as it exists in people. Yet you promoted Baker for his agression.
"What I am saying is a fact of history that is not stated on the sign.
"In order to be non-discriminatory, will the State of New Hampshire also affix the above statement on a sign placed under the existing sign? Better still, will the state remove their sign and rename the river, 'Lifeblood of the planet, that comes from the mountains.' Until it does, the State is reinforcing the prejudice against Nature that exists in people today. To have created a monument and named a stream after Baker is an act of bigotry that should cease. Its termination would be a way of improving your relationship with me."
Through the portrayal of our "history," we create importance and admiration for our environmental and social postures of today, prejudicially exalting ourselves to be a gift to the world as proven by our battles won, money made, distance from Nature (standard of living), treaties negotiated, and the life systems we have subdued or conquered. "History" helps to give us our myopic pride in being Americans, a feeling of nationalism that is unperturbed by our prejudice against Nature, or by the disrespect we reap from all we exploit. Ours is a history of a society that is in trouble because we are prejudiced against our own self-preservation.
Living and learning from Nature has shown me that much of that which we call history emanates from the south end of a bull headed north;

it's often not much more than nationalistic, anti-Nature bigotry, the state of an art in a society that reveres materialistic distance from Nature, instead of the life process. How else can we explain our historic fears and extermination of wolves, when on one hand there is no authenticated case of a healthy wolf attacking a person, and on the other hand, wolves have been known to adopt lost human infants and nurture them? Instead, why haven't we historically feared and exterminated the automobile? It is a historical fact that cars kill and maim hundreds of thousands of people annually. We don't exterminate cars, because history is not prejudiced against euphoric technologies like automobiles; it is prejudiced against Nature, thereby encouraging our extermination of Nature, and in the long run, of each other.

I have discovered that I myself, and practically everybody else, are living histories of our relationship to Nature. In other words, what we have historically imposed upon Nature, we also have imposed upon ourselves. How different is our polluted, cancerous future from that of the wolves' annihilation? Until our history courses and activities change this situation, the historic course of our relationship with the planet will continue to deteriorate.

It would seem best to teach whole life history by including the thoughtful examination of the history that our personalities, senses, and feelings carry and often convey. This process would teach us to actualize workable alternatives that would correct the mistakes of the past, instead of today's history process that mainly serves to reinforce our self-destructive prejudice against Nature. In an anti-Nature society, applying the Whole Life factor to history encourages the expression of an individual's self-preservation planet blue thoughts and feelings, and can offset the prejudicial myths that history conveys.

§

I don't believe I could be a student in a traditional school. Today when I do occasionally attend a class, I immediately and consistently bring up the WL factor whenever and wherever it should be discussed—which is all too often. I often become so disruptive to the class that I leave of my own accord, because after a while the one-man effort becomes enervating. It is vital that students anywhere and everywhere cooperatively find and follow the pioneering courage of their American forefathers and cope today with the vital frontier of self-preservation by confronting American education's deep rooted prejudice against Nature. How important it is that students insist that the WL factor be acknowledged and discussed by all persons in each and every situation where it pertains. Go through your textbooks in any subject, locate their anti-Nature bigotry, and demand that it not be taught in your school! That would begin to give Nature an equal voice and make school become appropriate education rather than cultural propaganda. It would make school more vital since the WL factor is the self-preservation in our cultural lives. (see page 224 for questions.)

The appropriate WL factor should be numerically determined and placed on every product and service of our society. There it would serve to educationally confront each of us with the real cost and meaning of our technologically excessive lives with respect to the self-preservation of ourselves and our children. There will be more on the practical application of this idea in later chapters.

Lame Deer:

One man's shrine is another man's cemetery, except that now a few white folks are also getting tired of having to look at this big paperweight curio. We can't get away from it. You could make a lovely mountain into a great paperweight, but can you make it into a wild mountain again? I don't think you have the knowhow for that.

The only other mountains carved up like Rushmore are some huge cliffs in Asia. They always show some Babylonian big cheese, or Egyptian pharoah, trampling some people underfoot, and the inscriptions always go like this: 'I, the great king, the king of kings, the living god, I smote fifty towns over there and buried the inhabitants alive, and I smashed fifty cities down there and had everybody impaled, and I conquered another fifty places on this side and had everybody in them burned up, and to show you what a big guy I am I had a thousand slaves carve up this mountain.'

It is the white man's arrogance and self-love, his disregard for nature which makes him desecrate one of our holy mountains with these oversized pale faces. It's symbolic, too, that this 'Shrine of Democracy,' these four faces, are up to their chins in rubble, a million tons of jagged, blasted, dynamited stone reaching all the way down to the visitors' center. If you look up the mountain, the way most tourists do, you see these four heads rising out of something like a gigantic, abandoned mine dump. But nobody seems to notice that.

We Americans will remain conflicted and prejudiced against Nature until we begin to cope with the knowledge that the planet is alive, and relate to it feelingfully with a love and reverence for life and with the respect that is due one's mother.

Sometimes the best way to observe how and why we relate to anything is to understand ourselves, our daily relationships and their origins. The human relationship with Nature should be found in books on human relationships, books on the origin of people, books on child development or social psychology—but it is not found there. Books on the social sciences are crammed with descriptions of people relating to people. There is almost nothing in them about people's relationship with Nature. Most psychology books would have you thinking that an individual's relationship with Nature just does not exist; that the human ego, not Nature, is the source of human feelings; that animals don't experience feelings as people do.

The basic relationships that sustain human life, relationships with Nature, are not even recognized, much less studied, by the experts in the field of human relationships.

That is what happens when you prejudicially decide that something is dead. You stop relating to it holistically, because you believe it won't interact, or you relegate it to another academic area so as not to be bothered by it. To blindly negate the existence of Nature as a part of everyday relationships from the time of birth is the epitome of prejudice against Nature and the planet.

Ecology is the study of interrelationships of life, including human life and geology. The science of ecology is terribly complex because Nature has had five billion years to evolve some highly intricate relationships. We are taught by experts that we need experts to help us understand Nature, for alone we can't understand life, and without help, we ignorantly injure our life relationship with the planet. Is survival that difficult? If so, how then is it achieved by all other life forms?

It is our prejudice against Nature and our internal nature that limits our understanding of and relationships with Nature. We demean our warm feelings for a seal and aggrandize a high-paying executive position. We feel alone and alienated because our prejudice has blinded us to the value of our natural feelings, the essential basis for our relationships and support. Much of our personal alienation and uneasiness with each other is due to missing the feelings, depth, symbolization, and stability with Nature. But when the WL factor is invoked, these symptoms subside, along with associated stress, depression, eating disorders and hostility. Why? Because the WLF connects the individual with the life-sustaining power of the planet's ongoing T-R relationships.

I first observed this connection while on an island solo. When the solo began, my mind was filled with problems I carried with me to the island. Minor concerns about conflicting past experiences and my expectations for the future engaged me while I sat alone on a rock overlooking the sparkling bay.

As I languidly stared at the engaging woodlands, mountains and clouds, a gentle breeze "told" me that I had been taking the atmosphere for granted. I was looking through it rather than including it as part of the scene, yet it was also a product of Nature's life-giving art. So I attempted to sense the air. I closed my eyes and again listened to the whispering wind, then to myself breathing, savoring each life-preserving interchange between myself and the planet. I held my breath; gradually I felt tension build as my body cried for oxygen. Finally, I could stand it no longer. My self-preservation feelings demanded that I inhale. I breathed deeply; the release of tension felt wonderful. Gone was my preoccupation with past and future problems. Instead I felt a powerful contact with my body and the planet. I sensed a unity, a vital identity with nature. We were the same and what was important was the present, the ongoing life relationship. I returned to the group and learned that other participants had gained some of the same fullfillment from their solos.

As we shared our experiences with each other, we recognized that the tension-release process is the universal pulse of the seasons, life systems, orbits, mountains, clouds, waves and tides. In people it produces the life-preservation feelings that had caused me to gasp for air.

By applying the WL factor to our solos we established a new awareness of Nature, especially with respect to time. It became apparent that nature and life only operate in the present moment. What we call time, the past and future, is a function of our mental images, symbols and feelings. They're so vivid that we often believe they're the actual past or future, and we relate to them as such. But they're not, nor do they contain Nature's T-R energies for these are only found in the present. An example is hitting a home run while playing baseball. Think of all the action, coordination, exercise, feelings, reflexes, noise, training, dirt, stress, excitement, energy, materials, people, places and sensations that go into a homerun. Few, if any of them are brought back or conveyed by saying, "Yesterday I hit a homerun," yet that statement is considered to be reality, an accurate portrayal of the past. It's an obvious distortion.

When we let our personal histories or expectations-our cubic redness-supercede our immediate relationships, conflicts are often produced by the differences between our inaccurate or misconstrued images and the present. In addition, our cubic-red conflicts produce stress unless they are modified by the T-R energies of the present. To illustrate: one student was constantly depressed about her unfulfilling relationships, weight and anxieties. She experienced a build-up of tension when she felt inadequate, but learned to enjoy the release she felt as she learned to talk about herself. Mountains became molehills as her involvement with her immediate thoughts, feelings and life drew her into the present and energized her ability to cope. Therapeutically, she connected with the T-R survival forces of the planet. Her depression disappeared as she enjoyed herself instead of fretting over the past or future. She said, "The greatest thing I got was a feeling of confidence, wholeness and belonging. With each rainbow I saw, each lizard I met, and every idea I expressed, I developed stronger feelings of security and trust in myself, people and the planet." During and following the school year, most of our expedition participants experience this kind of growth.

Undoubtedly, the heart of experiential education or therapy is their ability to help a person fully function in the present, in Nature's whole-life T-R energy field. When we don't invoke the present into our lives, we lose the healing powers of nature.*

Lame Deer:

You have even changed the animals, which are a part of us, part of the Great Spirit, changed them in a horrible way, so no one can recognize them. There is power in a buffalo—spiritual, magic power—but there is no power in an Angus, in a Hereford.

There is power in an antelope, but not in a goat or in a sheep, which holds still while you butcher it, which will eat your newspaper if you let it. There was great power in a wolf, even in a coyote. You have made him into a freak—a toy poodle, a Pekingese, a lap dog. You can't do much with a cat, which is like an Indian, unchangeable. So you fix it, alter it, declaw it, even cut its vocal cords so you can experiment on it in a laboratory without being disturbed by its cries.

A partridge, a grouse, a quail, a pheasant, you have made them into chickens, creatures that can't fly, that wear a kind of sunglasses so they won't peck each other's eyes out, **birds** with a **pecking order**. There are some farms where they breed chickens for breast meat. Those birds are kept in low cages, forced to be hunched over all the time, which makes the breast muscles very big. Soothing sounds, Muzak, are piped into these chicken hutches. One loud noise and the chickens go haywire, killing themselves by flying against the mesh of their cages. Having to spend all their lives stooped over makes an unnatural, crazy, no-good bird. It also makes unnatural, no-good human beings.

That's where you fooled yourselves. You have not only altered, declawed and malformed your winged and

and four-legged cousins; you have done it to yourselves. You have changed men into chairmen of
boards, into office workers, into timeclock punchers.

Is America prejudiced against Nature? Absolutely, and we will remain
prejudiced until we develop environments and symbols, and experience
feelings that allow us to react to the planet as if it were alive, with an equal
right to life. Nature's wholeness and integrity must be accepted as being just
as vital to us as is any mother to her child. We must use our culture to insure
survival instead of *to increase* our distance from Nature.

To subdue prejudice is to find it, confront it, and put the power of
change into the hands of those who are caught in it. Prejudice is people
stuck in destructive patterns of relating. Unlike the jargon of anthropology
or sociology, the concept of prejudice is not an unwieldy, overwhelming
conceptual scheme with a terminology that is available only to a few who
may understand it. Defining and confronting prejudice against Nature is a
positive step in helping people cope with problems. It demands that people
examine themselves and their life relationships in light of the Whole Life
factor. It is effective because it pertains and because it is accessible.

In order for the WL factor to deal with prejudice against Nature, it is
necessary not only for us to recognize our subconscious tendencies for
cultural euphoria and its adverse effects. We also must understand how
these tendencies are reinforced by our own institutions. We can't fully learn
this from our culture, for a bigot cannot properly teach equality. We can,
however, learn from our feelings and the planet itself.

*see Chapter 12 for further explorations of this phenomenon.

Chapter Eight

THE EARTH SHALL TEACH

The inconceivable trade-off begins as the minutes tick by. Where are the friends of the Earth; where are those Americans who say they care?

I have an intense desire to return to the womb. Anybody's.

—Woody Allen

Through binoculars, a pair of eyes eagerly scanned the gazelle herd and the surrounding countryside. Were the reports true? Was the three-year-old boy who had been lost in the wilderness seven years ago still alive? Had he really been found, nurtured, and raised by a wild herd of gazelles?

Suddenly the observer spotted what he had come so many miles to find. He wrote:

> From one of the bushes a slender white form springs, a gazelle, which immediately begins to pull up ball-shaped roots of dhanoun—the desert's principal survival food. Suddenly, I see blue flashes against jet-black hair, as a child with a bronzed and slender body darts from the same bush, throws himself at the unearthed roots, teeth first, peels them with clicks of his tongue, then cuts them up frantically with his incisors.
>
> My gazelle-boy makes for a seeping rock, sucks in the hesitant drops and then laps up those which fall, with a long pause between each drop as if to savor its taste. With a few leaps, the child heads for another corner of his oasis. I watch him sniffing almost incessantly, his neck stretched, his face to the wind, his nose continually jerking in little convulsive movements, his herd following behind him or leading him in turn. He also sniffs the flanks of his animals, sprigs of brushwood, bits of thorn-bush, flowers, berries, fallen dates, balls of dung, traces of urine. Dare I add that his nose also finds its way to his animals' hindquarters?[1]

Clearly, gazelle-boy would be considered a severely misguided and disturbed youth, if he entered our society and his peer group. He's a complete dropout: probably has an IQ of zero, with achievements to

103

match. He exhibits most of the immoral, uncivilized behavior and attitudes known to man. A dirty, rude nude who might bite old ladies, or pee on the grand piano—truly he would be a candidate for a padded cell or a cage at the local zoo. His only claim to fame might be that he did survive. He learned to survive in the natural world. Think about that, will you? Could you or I do that? Could we do it at three years old? Although we seldom recognize it, the environment is an incredibly successful teacher. It taught a human being to be like a gazelle. It also taught each of us to be what and who we are. The only difference is that our environment is different from the environment of gazelle-boy.

Studies have shown conclusively that the average American spends 95% of his or her life indoors: inside houses, cars, schoolrooms, offices, and bathrooms—inside the minor and major fortressess we have created to protect ourselves from the elements and wonders of Nature, including other human beings.

The indoor environment has powerfully conditioned us to it. We are conditioned to our man-made world as stringently as gazelle-boy was conditioned to the natural world. We know as little about relating to a peat bog as gazelle-boy knows about driving a car. Our ignorance is something to keep in mind as we make our decisions about how to exploit the natural world in order to maintain our standard of living.

Would you trust gazelle-boy to design and operate a corporate farm, O'Hare International airport, or Con-Edison? Just how much respect would you have for our economic and political leaders, if you thought they belonged in a zoo? Isn't that how Nature must perceive them?

Let me give you a specific example of the human condition. The Expedition was visiting the Louis Leakey Archaeological Site in Calico, California. The interpreter at the excavation explained how various artifacts there had been made. He demonstrated the use of a hammerstone to flake rocks and make a fist axe; then he invited the Expedition members to try our hands at it. For two hours, we used the rocks as hammers and pounded away at large pieces of chert, until they assumed the shape of fist axes or scrapers. It was very exciting.

That afternoon we traveled to Death Valley, arriving in a vicious dust storm. We decided to put up tents on the hard ground of the campsite, and proceeded to do so. What followed was startling.

> In blinding, choking dust, coughing, with eyes tearing,
> sixteen students informally lined up and waited their turn
> to use the solitary hammer in the tool box in order to
> pound their tent stakes into the hard ground. They com-
> plained that we should have had many more hammers
> with us to cope with this kind of situation. There were

rocks lying all around which easily could have served the same purpose. Only four hours earlier, we had been using rocks as hammers. But the unconscious conditioning of our culture prevailed, and the store-bought, steel hammer was sought, even at the cost of extreme discomfort from the dust storm. Obviously, dependence upon technology was alive and well.[2]

Through conditioning, not only is our red-cubic environment an authoritative teacher, but its symbols and images, which are internalized in the individual, often act in the same capacity. An encapsulization of this phenomenon, based on scientific conjecture, is found in the following allegory:

> The black hole exploded once again, slinging out into space the cosmic dust which makes up the universe. Quarks, neutrinos and hundreds of other atomic particles flowed and reacted to create the night sky. All was in motion, and remained in motion. The energy flow took different courses, producing different forms which interacted with each other. New forms were created, each with an integrity of its own, each integrated with the total energy flow.
>
> Near the rim of one galaxy of energy was an inconspicuous star around which orbited planets and asteroids. On the third planet from the star, the energy field warped in an unusual fashion for the area, creating a special combination of atoms and energy that was able to experience life and enjoy maintaining it. In the latter days, the compound received the name protoplasm. The energy which formed the universe continually formed protoplasm into new life forms, experiments which supported and depended on each other and had protoplasmic qualities. A great chain of life evolved: each organism sensitive, each sustained by the other, each attempting to survive in the milieu, each exhibiting the life process, be they of one cell or many.
>
> Red Dog and Grey Dog were links in the three billion year old chain. They had just arrived in a new area where rabbits abounded. They moved into a leaky cave which gave them minimal protection from the elements of nature. When it rained, they got wet. When it was warm,

they were warm. Whatever happened in the environment happened to them, but not enough to deprive them completely of their life energies. With minor adaptations, they survived, but always under the tension release pulse of the environment. A tighter cave reduced the discomfort of wetness and chance of death, but death was always present, always attempting to make energy for others from the energy of the two dogs. That was the name of the game of life. Discomfort was a signal that death was increasing its pressure. Vigilance was the watchword for all. Anything providing shelter or energy from the environment was treasured.

One fine day the two dogs went hunting and chanced upon a rabbit which ran through some cactus in order to escape. Both dogs ran into the spines and yelped with pain, but sped on. The next week, the same incident was repeated in a different area. That was how the rabbit survived. The cactus was its protector. After many months, the dogs had run into and located many of the cactus plants. Those they repeatedly ran into they learned to avoid.

Food became scarce in the immediate area and again the dogs moved. There were new cacti in the new area and the repetitive learning process started again.

Nobody can be sure exactly how the change took place. Perhaps it was a mutation, perhaps it was from selection or inbreeding. But something had changed. For when the dogs returned from the hunt, Red Dog said, "Grey Dog, you don't get spines in you anymore. How do you avoid them?"

"I noticed that," said Grey Dog. "I don't rightly know how to convey it. All I can tell you is that I can see the plant that has the spines, I can recognize it by its smell, I know which plant has the spines. I can see it right now, and I avoid it. It helps me catch those rabbits, too, you'll notice. I'm way ahead of you."

"Wait a minute, hold on, did you just say you can see it right now?" said Red Dog. "You're out of your mind, there isn't a spiny plant anywhere in this cave."

"That's just the point," said Grey Dog. "I can see it anyhow. I can see it behind my eyes even though I can see through my eyes that it isn't there. I have no idea how it happens, but I see it, or an image of it. I carry the image around in my head. When I see a plant that looks like the image, I avoid it, and end up with rabbits instead of spines. Let me tell you, it's so fantastic and so rewarding, that I find I rely as much as I can on my use of the image of the plant. It avoids discomfort, it provides pleasure. Why shouldn't I do it."

"Can you teach me how to obtain images?" said Red Dog.

"I'd like to, I really would," replied Grey Dog, "but I don't even know how I do it myself. It just happens, it's a gift, I was given a gift, but I don't know how it's made. All I know is that I can relate and respond to the internal image of the spiny plant, and it helps me survive better than before. I can expend less energy to obtain more energy from the environment. Boy, am I lucky."

A few days later, both dogs returned from the hunt. Grey Dog had gotten the rabbit, but Red Dog had no spines in him.

"Hey," said Grey Dog, "now it's happening to you. You have no spines in you."

Red Dog looked happy and replied, "Yes, it's happened to me but in a different way. I'm following you. When you avoid a plant, so do I. So now I, too, don't get spines."

"But you don't get rabbits either," said Grey Dog, "although you do help by limiting their choice of where they can go."

"I'm happy with it if you are," said Red Dog. "I'm avoiding those spines and eating better because together we're catching more than ever. It's terrific!"

"A few new things have happened with me," said Grey Dog. "I've met some other dogs who can do the same thing I can do. We're cousins. We all had the same grandparents. We've discovered that not only do we have images in our heads, but we can convey these images. I can describe the spiny plant and put the picture behind my cousin's eyes even though he never saw a spiny plant. I can draw a picture of the plant, and he can get the image. I can tell him to avoid it, and he sometimes can do that. And we're able to produce images of practically everything we can sense. We can remember them. We call it symbolizing. We relate to the images as if they were real.

It works so well that at times I find I'm relating to the symbol and think it's real, and vice versa. It gets confusing."

"Gee," said Red Dog, "I've met a lot of dogs who can't do that. They're just about starving cause you guys are getting all the rabbits. I'm sure glad I'm your friend and we get along. Otherwise, I'd be full of spines and hungry, too."

Many years passed and many things changed. One day, Grey Dog returned to the cave after playing chess with his cousins. He had no need to hunt that day because he had used an idea from his cousin and had borrowed a rabbit from him so he could play chess for the day. When he entered the cave, he could not find the rabbit he had saved. "Hey, Red Dog," he said, "have you seen the rabbit I saved for today?"

"Saved? What's saved? What does that word mean? Each time you come home from your cousins, you have some new thing. I don't know what in the name of Rin-Tin-Tin you're talking about."

"Where is the rabbit I left here?" demanded Grey Dog, sharply.

"Oh, that rabbit," said Red Dog. "Don't worry about it. I ate it this morning after you were gone."

"You stole my rabbit," said Grey Dog. "You took it without my permission."

"What the hell are you saying," said Red Dog. "Stole? Permission? What does that mean? I never heard of them."

"Bad dog, Bad dog," said Grey Dog and he charged at Red Dog.

"Hey, Grey Dog, don't bite me," said Red Dog. "I can see you're angry, and I don't know why. But I do know what's going on, because this is not the first time this kind of thing has happened. Last week you came back with the idea that I was bad because I disobeyed you, whatever that means. The week before I didn't respect you, whatever that means. Before that, I was a bad dog because I ran into a spiny plant and you couldn't get the rabbit because you were depending on me to be up with you and to scare the rabbit your way. Before that I was a dumb dog—stupid, because I didn't obey the rules or whatever.

What rules are, I don't know. Each time you're angry and try to punish me you bite, you push me in the spiny plant which you now call a cactus. You've just about used up the inept part of the rabbit population; the rabbits that are left are much harder to catch, so you're not even getting food with less energy—you're back where you started. And you're always angry or unhappy about something. You're not even as happy as in the good old days. And I know why, too, because I've been watching it happen right in front of my eyes. It's downright disgusting."

"Red Dog, I'm not sure I will tolerate this insolence. Tell me what you've seen. Stop keeping secrets from me. That's wrong, you're not even honest. Bad Dog, Bad Dog."

"Insolence? Secrets? Wrong? What does that mean?"

Grey Dog grabbed Red Dog's neck in his jaws.

"O.K.," said Red Dog. "I'll tell you what's going on. You think you're different than me—that you've changed. But you haven't changed one little bit except that you can symbolize. That's the only change."

"Out with it, why am I so unhappy?"

"Because you're scared—afraid—terrified of what's inside you. You're so scared you won't even let yourself see it. That's why you have to ask me now. Otherwise, when it shows up, you automatically switch to other, more pleasant symbols, even though you still feel uncomfortable, and are therefore obnoxious and anxious."

"What is inside me that is so terrifying, so upsetting?" asked Grey Dog, scowling.

"Your feelings! Your feelings about discomfort and death in the environment. You're at their mercy. Oh, that's what's doing it all right. When you were first able to avoid a cactus by avoiding its image behind your eyes, it was obvious to me that the pain, the discomfort of the cactus spines was also behind your eyes. But you could switch it off by thinking, dreaming, fantasizing that you could avoid the cactus, avoid the pain. You've been doing it ever since. You've conditioned yourself. You conditioned yourself into believing, knowing, that you could avoid the feelings of death and discomfort by manipulating their images in your mind and shutting them off. You fantasized rules, you fantasized that the things which would possibly let your hidden, hurtful images appear were bad, wrong, distasteful. To escape the discomfort of nature within, you've created schemes of good and bad, right and wrong. Bad things are anything which makes you feel your hurtful images. You now live by the rules of these hurt-avoiding schemes rather than by your feelings from nature. You think you've escaped the fluctuations of nature by making your real environment conform to your schemes. You've swapped a world full of rabbits for their images. But you can't eat images and you know it. You've got the tensions of nature burning within you, and rather than realize it, you're beginning to destroy your environment because you think its hurtful. That's why you're unhappy, Grey Dog. It's actually the unconscious images inside which are hurtful. You're living in a fantasy world full of distorted anxieties, fears, misapprehensions, rules, tensions, punishments, and ignorance. It hasn't happened to me because I can't symbolize—but I'm doing OK. Members of the great chain of life can't symbolize...but they're doing OK. But they were better off before you were able to symbolize and utilize them as dirt to extinguish your fantasy world fires. Just yesterday you were chewing down bushes and bringing them into the cave so you could lie on them to protect yourself from your inner fear, your fantasy, your image of being cold. I sleep in the same cave, I feel the cold of nature, I find it nice to know nature is still there, 'cause that's where I get my food. That cold is a comfort because I don't know anything else. But you've killed bushes to protect yourself from something which is harmless. Will nature ever forgive you?"

"Are you finished?" asked Grey Dog harshly, keeping his hold on Red Dog's neck.

"Just one more thing. The greatest hoax which you play on yourself and the environment is the distortion that your images are accurate reflections of the environment. It's a cruel hoax. You never realize that your image is as false as water is dry. A cactus plant is a lot more than just a hurtful pinprick. Cactus is what protects the rabbit from getting caught, so there are still rabbits around to be eaten. The wild pig eats the cactus. His feces fertilize the grass, his diggings aerate the soil so grass can grow, the rabbits eat the grass and the rabbits are our food. The cactus plant is part of our survival. Is that part of your image? No, not at all. There are hundreds of other things going on with a cactus plant. Are they all part of your image? No, you know they're not. You're not even aware of them...I've already heard you say to your cousin that it would be great to somehow remove all the cactus plants so the rabbits couldn't hide and you wouldn't get stuck. You were also talking about destroying an aspect of your life which keeps you alive—your environment—just because your image lies to you."

Red Dog's discourse had inadvertently activated Grey Dog's unconsciously learned inner fears of being hurt by being wrong, rejected. Grey Dog could feel them, but they appeared as Red Dog being obnoxious.

"Red Dog, roll over!" barked Grey Dog furiously.

Red Dog rolled over. He lay there, thinking, "What the hell am I doing here on my back at Grey Dog's feet and command? Rats! He's been part of my environment and has conditioned me. And I've become dependent on him. Screw this. When I get up, I'm leaving, only because I don't know how to do something about being conditioned. I don't have the tools. Now, if I could symbolize, I could fight conditioning tooth and nail and still live with nature. I'd use my ability to symbolize wisely. I'd use it to maintain myself by maintaining my environment, and fighting irrational and ignorant, false images through awareness."

Today, casual observation of dogs, wolves, coyotes, and their genus, reveals that Grey Dog is no longer to be found. He somehow became extinct—a victim of some devastating aspect of his environment. But the ability to symbolize again appeared in the great life chain, at a later date, this time in an anthropoid ape. It was assisted by this ape's ability to pick things up with an opposable thumb, to verbalize, and to make and use tools to increase its power over the environment. One need only look in a mirror to see that anthropoid today. What do you see? Yourself? No, that's only an image of yourself which you imagine to be yourself.

What is your self image? Grey Dog? Red Dog? What do

you do with the uncomfortable feelings you carry around inside you? Which dog will you be most happy being?

As the allegory indicates, you and I live in two environments simultaneously: the *real world,* and our images of that world. Each environment can act as a conditioning stimulus for us. As my years on the Expedition and in Nature grew in number, I began to recognize, on a feeling level, what I had already known intellectually: that through conditioning, as a kid, my accumulated thoughts, actions, and feelings had become for the most part habitual. The rewards or punishments of *society,* (my family, relatives, teachers, and government) had steered me in culturally acceptable directions—seldom asking me whether these were the directions that I desired for myself. I had been re-educated from Nature in a manner similar to the re-education of the Hopi woman we visited. Society rewarded me with various aspects of tension-release. I received reinforcements of my womb-euphoria feelings (like being hugged and praised), necessities for survival (my food, clothing, and home), and relief from discomforts (helping me with relationship problems or taking care of me when I was ill). My *punishments* included disapproval, rejection, and occasionally, corporal punishment, from parents, teachers, and friends. Each act, thought, or feeling that gained me one or more rewards or punishments became either stronger or weaker, as gratification or discomfort was experienced. Repetition of the process made many of my thoughts, feelings, and actions become set, automatic, and unconscious. Their general direction was to move away from Nature. For example, in a restaurant I recently witnessed a mother take away a plate of food from a crying young child saying, "If you don't wear your bib, you can't have your dinner!"

On the Expedition, I discovered that I had forgotten some of the rewards and discomforts that made me act and feel the way I did; I was left glued to red-cubic behaviors, thoughts, and feelings that I could not immediately explain or direct. They were ingrained. Together they made up my individual upbringing and conditioning. For example, on being exposed to my favorite foods, I would repeatedly eat too much. I was manipulated by my chocolate cake environment as thoroughly as one who goes to a party and is turned on to smoking and drinking, even though he or she *gave it up* three weeks before.

I discovered that I was marching to a cultural authority that had been superimposed upon my natural self-preservation feelings. Like the Hopi woman we visited, subconsciously, there was a child inside of me who was not happy with the falsehoods he had been conditioned to believe about Nature and himself.

I began to notice that even informal exposure to an environment could begin the conditioning process. Contact with people that used different speech patterns sometimes led me to unconsciously use those patterns.

The Expedition met a graduate student who told us about an interesting experiment which demonstrated the conditioning process. She said that meaningless Chinese-type figures of calligraphy were shown to a group of subjects. Some figures were shown only once, others as many as twenty-five times. The subjects were then asked which figures they liked. They consistently chose the figures to which they'd had the most exposure, followed by those to which they'd had only slightly less exposure. The figures to which they'd had least exposure were least appreciated. Why did this occur? There are probably many ways of explaining it, but my thought is that it happened because of recognition of the repeatedly exposed figures had been most rewarded. Life is rewarding; it feels good to be alive. The unconscious reward, the joy of living, accompanied each interaction and was reinforced with repetition.

In looking back over the years, I can identify four major conditioning agents that have reinforced my prejudice against Nature: money, religion, rejection, and social pressure. Lame Deer's thoughts about these subjects provide an interesting contrast to my own.

Lame Deer:

The bald eagle is your symbol. You see him on your money, but your money is killing him. When people start killing off their own symbols they are in a bad way.

The Sioux have a name for white men. They call them **wasicun**—fat-takers. It is a good name, because you have taken the fat of the land. But it does not seem to agree with you. Right now you don't look so healthy—overweight, yes, but not healthy. Americans are bred like stuffed geese—to be consumers, not human beings. The moment they stop consuming and buying, this frog-skin world has no more use for them. They have become frogs themselves. Some cruel child has stuffed a cigar into their mouths and they have to keep puffing and puffing until they explode. Fat-taking is a bad thing, even for the taker. It is especially bad for Indians who are forced to live in this frog-skin world which they did not make and for which they have no use.

Have you ever wondered why you may feel so strongly about money, and why people are driven to obtain more money than they need for survival? I have heard that money evolved to facilitate people's ability to exchange goods. My own experiences tell me that it has also become a powerful reward. I have come to realize that from my early childhood to the grave, every time that I spend money, some need of mine is met, some drive is satisfied, some relationship is controlled, some prejudice is reinforced. I grew up conditioned to money.

Through conditioning, overpowering life-long feelings have been transferred to money. For me at one time, money was a conditioned addiction; I was hooked by the most persuasive (and pervasive) addiction in America.

In the past, money bought me technological security, my cultural release from tensions of Nature. I was enamored with and goaded into the acquisition of *new* technologies and, therefore, usually desired more money. I found that money could buy my escapes from uncontrollable relationships and uncomfortable circumstances. I found myself enticed by price discounts to purchase needless and superfluous technologies. As I gained professional status, I found that it cost me more to maintain my status. I was usually in debt no matter my income.

Our civilization is fueled by money. Some people may cheat, steal, manipulate, exploit, or even kill for money. My friends and family and I designed our lives around the dollar—it was unknowingly our metaphysical symbol.

I was helplessly caught up in this web of our civilization's making. I believe all of us are in the power of those who have the money to buy our relationships, our attitudes, and our environment. At one time, money was the reward that prodded me into unrewarding, distasteful, and harmful work. In visiting and revisiting America, all too often, I have observed that money is the steering mechanism of the juggernaut that converts Nature into natural resources. The desire for money and the technologies it buys has lured me and you from harmonious relationships with Nature; it has destroyed equilibrium between us and the planet. Our love for life has been acculturated into *I love to shop.*

Lame Deer:

The piercing could be done in four different ways. For the **Gazing at the Buffalo** way the flesh on the dancer's back was pierced with skewers. From these were hung up to eight buffalo skulls. Their weight pulled the skewers through the flesh after a few hours.

The second way was **Gazing at the Sun Leaning.**
This was the one most used. The flesh on the dancer's
breast was pierced about a hand's width above each
nipple and a wooden stick or eagle's claw stuck right
through the muscle. At the end of the dance each man
had to tear himself loose.

Some white men shudder when I tell them these
things, yet the idea of enduring pain so that others may
live should not strike you as strange. Do you not in
your churches pray to one who is **pierced,** nailed to a
cross for the sake of his people? No Indian ever called
a white man uncivilized for his beliefs or forbade him
to worship as he pleased.

The difference between the white man and us is
this: You believe in the redeeming powers of suffering,
if this suffering was done by somebody else, far away,
two thousand years ago. We believe that it is up to
every one of us to help each other, even through the
pain of our bodies. Pain to us is not **abstract,** but very
real. We do not lay this burden onto our god, nor do we
want to miss being face to face with the spirit power. It
is when we are fasting on the hilltop, or tearing our
flesh at the sun dance, that we experience the sudden
insight, come closest to the mind of the Great Spirit.
Insight does not come cheaply, and we want no angel
or saint to gain it for us and give it to us secondhand.

Like money, religions create culturally induced euphoria through the
conditioning process. Baptists beget Baptists, just as Democrats beget
Democrats. I was not born into a family that practiced the rituals of
organized religion, but I have lived with many people who believe that the
rewards of heaven (supported by communion, graces, group chanting of
prayers, the hope of womb-euphoric spiritual exhaltation in a *life-
everlasting*) are countered by the punishments of hell (reinforced by shun-
ning, excommunication, the threat of damnation). I have seen the depth of
those beliefs and how they can act as powerful, guilt-producing conditioning agents in religious groups.

I have lived with people whose religious philosophies promise the
reward of God's approval to those who *subdue* the Earth, embrace the
work ethic, and pursue material wealth. I have seen the effects of religious
institutions which reward religious belief systems that are prejudiced
against Nature. Learning that we are born *sinners* makes us susceptible to a

need for approval from religious groups and their potential anti-Nature attitudes. Our love for life has been conditioned into *I love those who forgive me for being natural*.

Lame Deer:

> Americans want to have everything sanitized. No smells! Not even the good, natural man and woman smell. Take away the smell from under the armpits, from your skin. Rub it out, and then spray or dab some non-human odor on yourself, stuff you can spend a lot of money on, ten dollars an ounce, so you know this has to smell good. **B.O.,** bad breath, **Intimate Female Odor Spray**—I see it all on TV. Soon you'll breed people without body openings.
>
> I think white people are so afraid of the world they created that they don't want to see, feel, smell, or hear it. The feeling of rain and snow on your face, being numbed by an icy wind and thawing out before a smoking fire, coming out of a hot sweat bath and plunging into a cold stream, these things make you feel alive, but you don't want them anymore. Living in boxes which shut out the heat of the summer and the chill of the winter, living inside a body that no longer has a scent, hearing the noise from the hi-fi instead of listening to the sounds of nature, watching some actor on TV having a make-believe experience when you no longer experience anything for yourself, eating food without taste—that's your way. It's no good.

Rejection is discomforting, because it triggers anxious feelings that unconsciously tell us that we have no niche. These feelings may originate from the birth experience where we are *rejected* by the womb and forced into the tension of Nature's fluctuations. In the Expedition community, we see many examples of how the withdrawal of emotional support is often perceived as a threat to survival, especially in young people.

Rejection is a form of punishment. Because it can be so uncomfortable, it can be a strong conditioning agent. I can well remember adults, teachers, institutions, and other authorities entrusted with the education of children, that rejected me, and other young people when we expressed our natural feelings and interests. We were *dirty* or *bad*.

We Americans abhor Nature in ourselves. Feelings, blood, mucus, sweat, saliva, urine, phlegm, feces, pus, ear wax, sex, and sex organs, regurgitation, natural odors, nudity—all are natural; and to greater or lesser extents, all are taboo in our culture. Cleanliness is next to godliness. I was encouraged to treasure shiny sterile places, thoughts, and feelings, that created distance from Nature. Four-letter swear words were taboo because they symbolized Nature. When I watch television commercials today, I note that, with few exceptions, gaining distance from Nature is their prime pitch—no matter the product. Our love for life has been conditioned into *I love to feel clean.*

Lame Deer:

Why do Indians drink? They drink to forget the great days when this land was ours and when it was beautiful, without highways, billboards, fences and factories. They try to forget pitiful shacks and rusting trailers which are their **homes.** They try to forget that they are treated like children, not like grown-up people.

We drink to forget that there is nothing worthwhile for a man to do, nothing that would bring honor or make him feel good inside. There are only a handful of jobs for a few thousand people. These are all Government jobs, tribal or federal. You have to be a good house Indian, an Uncle Tomahawk, a real apple—red outside, white inside—to get a job like this. You have to behave yourself, and never talk back, to keep it. If you have such a job, you drink to forget what kind of person it has made you. If you don't have it, you drink because there's nothing to look forward to but a few weeks of spud-picking, if you are lucky. You drink because you don't live; you just exist. That may be enough for some people; it's not enough for us.

Economic and peer pressures were the conditioning agents that, when I was young, urged me to conform to the standards and styles of my culture. I was pressured toward being *cool* and gaining cultural euphoria from fashions, technologies, chemical intake, work, and other economically feasible euphoric stimulants. I was rejected in some circles when I was not *cool*—in those days, to be rejected was known as being *square,* or *out of it.* Like others, I was extremely sensitive to such pressures and, through repetition, I was conditioned to desire the security of social status. (Still, in many situations, I chose to be *square;* the decision earned me approval from my family and the Boy Scouts.) Our love for life has been conditioned into *I love to be accepted.*

On the Expedition, it became clear that each of these forces had reinforced and further ingrained my euphoric anti-Nature attitudes and feelings, far beyond the intensity needed for survival. Disguised as *security, acceptance,* or *health,* I had unknowingly learned that gaining distance from Nature and people was of cultural value. I noted that in our civilization, when gaining distance was stymied, womb euphoria was often obtained chemically through alcohol, drugs, and cigarettes, or in sexual release or non-participatory entertainment such as television. I could not help but notice that each of these destructive activities support major American industries which, in turn, encourage dependence on them. Today I believe we drink alcohol for the same reason that Lame Deer says an Indian drinks—to gain a temporary euphoria by dulling the pain of our distance from Nature.

The Expedition has helped me to see that cultural conditioning maintains the pressure and prejudice against Nature that begins at birth. Through this conditioning, the symbols of technology, money, and religion become powerful kings and queens.

I have found that the traditional educational system does more to convey and reinforce these conditioning agents than it does to negate them. The system itself is subject to these pressures, and its primary purpose is to teach students the skills that will make them successful in the culture, and provide them with distance from Nature.

By examining my own experiences and those of gazelle-boy, I've discovered that the immediate environment is a powerful educator. If an environment is prejudiced against Nature, so are the people within it. The Grey Dog and Red Dog allegory has illustrated the impact of imagery and symbolization. If a person's anti-Nature symbols and images are allowed to develop, that person will be prejudiced against Nature. The conditioning process reinforces the prejudiced womb euphoria tendencies that develop during early childhood.

Each conditioning agent helps to establish a person's prejudice. Operants of culture are used to humanize the infant, who arrives as an immature mammal and is transformed into a personality adapted to life in the symbolic, cultural world of its parents. As children grow, they learn to participate in their parents' social order. This transformation, known as acculturation, takes place in all human groups. It operates to perpetuate traditions through each succeeding generation and to maintain historically developed designs for living. Acculturation, therefore, will convey prejudice against Nature, unless it is confronted and creatively modified to encourage harmony with Nature.

All over the world, despite wide differences in climate, landscape, flora, and fauna, Nature operates through the

same physical, chemical and biological events and processes....Yet each group of people has established its own
cultural world, a world that it has created by transform-
ing the actual environment into a symbolic world of
meanings and goal values. These symbolic worlds are
products of the selective awareness and patterned perception of those dimensions in Nature of which each group is
aware and which it believes to be significant, if not crucial, for its existence.

While existing in the geographical world of Nature and
exposed to its varied impacts and opportunities, each
cultural group has created a symbolic world of meanings
which it interposes between itself and Nature. Thus a
cultural world arises as a transformation of the actual
world into an *as if* world.

To live in the *as if* cultural world, every member of the
group must learn to perceive the actual world and to act
according to its symbolic meanings and purposes
implied.

Each cultural group has developed its own ways of
relating to and communicating with Nature....

Thus each group, guided by its beliefs and assumptions
about Nature and human nature as imaginatively por-
trayed by its poets, prophets, artists, and more recently
by its scientists, strives to maintain its cherished ways of
living by shaping, molding, attempting to fit the newly
born infant, the child, the adolescent into the kind of
personality the group believes to be desirable, if not
essential, to its continuance as a people, requiring him to
learn its ways of living and conducting his life career.[4]

In these often unrecognized ways, *our* culture reinforces our early
childhood prejudice against Nature, leaving the individual to experience
the consequences.

Although I did not recognize it at the time, this was my situation in
1945, when at the age of 15, a rude unkindness jolted my perceptions of my
culture's relationship with Nature. The blow came from Walter Miller,
who, during a gym class soccer game, yelled to his friends, "Let's get the
little jewboy,"and followed up with a hard knee to my groin. Walter was
almost a foot taller than I and 50 pounds heavier. His kick left me reeling,
and with a problem as well. Why did this happen, and what could I do
about it? My friends told me it had something to do with "history," that the

Jews killed Christ, but that didn't help much. Physical resistance on my part was impossible without receiving more of the same treatment from Walter and the lot of his friends.

With the painful conflict in my mind and elsewhere, I made my way to my biology class, where within the hour, I fantasized that by studying *life sciences* I could scientifically gain control of the factors which control life, control people, and thereby control Walter Miller and his nasty friends. Unknown to anyone, that incredible distortion became a major motivation behind eight years of further misguided high school, college, and graduate school studies in the life sciences. Only after I received my M.A., did I fully realize that biological science had provided few solutions to the problem of prejudice that had originally brought me into it. A better solution might have been to achieve a Black Belt in Karate. However, my education did give me the insight that prejudice was a social or cultural problem while Nature, on the other hand, was fair; Nature appeared to be equally "cruel" or "supportive" to every living thing.

There followed years of studying guidance, the army, teaching and directing summer camps and travel programs. These brought me to the year of 1959 and the beginning of a self-founded career of expedition camping-out and firsthand observations of Nature and people. Only then, as I was living outdoors, did the planet have its chance to expose and explain to me its side of its ongoing argument with our prejudicial society.

I began to hear Nature.

I listened to the Labrador winds howl at 75 MPH while Diana, Timber, and I lay snug and safe behind a few stunted alder bushes that Nature had considerately provided. We were thankful for Nature's thoughtfulness on the tundra.

In the snowy Adirondacks, we felt the temperature drop to -18° F, and yet we remained warm by wisely conserving the body heat with which Nature had endowed us.

Time after time, year after year, Nature's "threats" to our survival were comfortably thwarted by natural means. I began to understand that for every survival problem that Nature produced, she also provided remedies nearby for those who cared to seek them. It became apparent that by fully using the T-R process, life was able to harmoniously sustain itself.

With further time spent in the out-of-doors, it became impossible for me to ignore that all my material possesions emanated from the planet, that indeed I, my consciousness, and my feelings also were a product of the Earth. Again and again my social and survival needs were happily met by Nature in ways that were so individualized that one would have to experience them to understand their specific meaning to one's self.

I remember the day that I sat staring at a single portion of a magnificent waterfall in the Grand Canyon, and after a while, shifted my eyes to the adjacent red rocks. By magic they appeared to rise, as if they were a fluid defying gravity, an impossible waterfall of rocks flowing upward. I was shaken and thought I was hallucinating, until the motion of the rocks subsided. Curious, I repeated the sequence and obtained the same results (which I urge the reader to experience at your own local waterfall by staring at a single point in the falls, letting the water flow through that point). Perhaps this was Nature's way of teaching me that what I actually saw at any given moment could be seriously influenced by my previous experiences.

The phenomenon of the rising rocks led me to question the validity of my daily perceptions. Was what I saw at any given moment reality, or was it a view of the world that was biased by what I had been taught and by the process by which I had learned it? As time went on, it became obvious to me that many of the answers that I had been previously fed at home and school might not have been the wisest or only answers. It was these misguided answers that no doubt had led Walter Miller to hating Jews. His was not an act of Nature.

Nature and the planet's role in the survival process became so real, active, and nurturing, that they both had to be considered to be alive, to be a living mirror image that made my life possible. They were the mold that formed the cast. From somewhere came the thought that with the exception of my cultural upbringing, I and Nature were exactly the same thing. Walter Miller maliciously kicking my genitals was a culturally induced prejudicial act against Nature, for indeed my reproductive glands and their pain were pure and simple Nature. In a strange, marvelous way, by applying the Whole Life factor, a 35-year chain of events began to make sense, as I recognized the incredibly prejudicial manner in which I and other folks had been taught to experience and relate to Nature. Upon further observation, the effects and extent of our society's prejudice against Nature slowly became clear as did its reversal in students who applied the Whole Life factor to their relationships with Nature and people.

§

I have noticed that in Nature, wherever the strong winds blow, the trees collectively maintain the same height, thus mutually protecting each other by providing the least amount of tension from the wind, snow, or salt spray. Occasionally one tree grows higher than the rest—as if it were cooperatively supported by the others to pioneer the stormy space above the tree colony. This one taller tree, if it survives, becomes a new barrier to the wind, a shelter that might assist the growth of leeward trees. Perhaps in this way, the wind-stunted grove of trees experiences a trifle less tension, gains a better foothold on life, and a new strength for relating to the wind and the planet.

In a similar way, the concepts of prejudice against Nature, the Whole Life factor, and our distance from Nature introduce a relatively new dimension to America's collective conscience. This dimension has met with strong resistance from most of the publishing world for reasons that I have already described, yet I find that it offers new areas of personal and environmental growth for those who are concerned about self-preservation, harmony, and a more globally responsible way of life. The concept gives Nature a voice with which to pioneer its way into our presently existing consciousness.

REFERENCES

[1]ARMEN, JEAN CLAUDE. *Gazelle Boy*. Universe Books, 1974.

[2]COHEN, MICHAEL J. The Equilibrium of Survival. *Environmental Education Report*. Washington, D.C.: Center for Environmental Education, October, 1981.

[3]COHEN, MICHAEL J. *Across the Running Tide*. Cobblesmith, 1978.

[4]FRANK, LAWRENCE. *in* Falkner, Frank. *Human Development*. W.B. Saunders & Co., 1966.

Summary and Conclusions

Be it the wilderness, a gazelle herd, society or our symbols and imagery, the environment that surrounds our inborn nature conditions it. Through education, reinforcements, punishments and repeated exposure, the people, materials and prejudices of our civilization acculturate our natural thoughts, feelings and actions. Dominant mores such as economics, religion, cleanliness and acceptance by authorities become powerful habits. We learn from our culture that it and our habits are good while Nature's T-R fluctuations are to be avoided. Distance from, or power over Nature's pulse are rewarded by social status, security, and survival. Mother Earth's polluted environment represents her inability to deal with our misguided mentality.

Chapter Nine

BEHOLD A PALE HORSE

Cultural excesses provide satisfactions for those who are unable to find fulfillment in the pulse of people, Nature, or themselves.

There is a story about a professor who decided to test the ability of a frog to jump. The frog was placed in the center of a bull's-eye circle. The little professor would violently wave his hands and yell, "Jump, frog, jump!" at the top of his lungs. The frightened frog would jump and the distance of the jump would be measured. The professor then removed the right front leg of the frog, and, placing the frog back in the circle, once again yelled, "Jump, frog, jump!" Again the frog jumped and the distance of the jump was measured. It was noted that the frog did not jump as far with one leg missing. The left front leg was next removed and the procedure repeated with the result that the frog jumped a shorter distance with two legs missing. Again, the procedure was repeated, removing first one, then the other, of the frog's rear legs. Finally the completely dismembered frog was placed in the center of the circle and the command was given, "Jump, frog, jump!" To no avail. The frog did not move. "Jump, frog, jump!" boomed the professor, even louder. Nothing happened. The professor clapped his hands sharply and screamed, "Jump, frog, jump!" The dismembered frog remained motionless. The professor wrote in his report, "A frog with four legs missing is deaf."

The significance of this story lies not only in the error made by the professor, but also in the realization that we actually recognize all of the components of the story: science and academia unnecessarily experimenting with live animals; scientists drawing inaccurate conclusions from unholistic inquiries; academicians dissecting a whole living creature into parts in the belief that this will tell us how that creature functions. We know that there exist scientists callous enough to be party to such an experiment—human beings who could attempt to frighten a dismembered frog or an exploited person and remain untouched by its plight. America's prejudicial relationship with all of Nature is epitomized by the professor's relationship with the helpless frog. A further significance of the story is that practically everything in it is a reality of our civilized life. The components of the story are components of me and of you.

§

A recent environmental impact report stated that a proposed power dam was an asset to wildlife because it would assist the survival of a local Bald Eagle population. The dam's turbines would stun the fish that came through them, thereby making them more accessible to the eagles. Is that twisted, prejudicial thinking any different from the promotion of a bass fishing contest where $5,000 in prizes brought thousands of people to take the lives of a wild fish population for money? Is it different from referring to wild animals as *game?* What is the "game" except prejudice against Nature, a game that is overplayed by those people who mistakenly seek womb euphoria from technology and must forcibly extract their euphoria, the womb of life? Are the effects of this game destructive? "But ask now the beasts and they shall teach thee; and the fishes of the sea shall declare unto thee." JOB 13:7-8

Lame Deer:

You don't want the bird. You don't have the courage to kill honestly—cut off the chicken's head, pluck it and gut it—no, you don't want this anymore. So it all comes in a neat plastic bag, all cut up, ready to eat, with no taste and no guilt. Your mink and seal coats, you don't want to know about the blood and pain which went into making them. Your idea of war—sit in an airplane, way above the clouds, press a button, drop the bombs, and never look below the clouds—that's the odorless, guiltless, sanitized way.

When we killed a buffalo, we knew what we were doing. We apologized to his spirit, tried to make him understand why we did it, honoring with a prayer the bones of those who gave their flesh to keep us alive, praying for their return, praying for the life of our brothers, the buffalo nation, as well as for our own people. You wouldn't understand this and that's why we had the Washita Massacre, the Sand Creek Massacre, the dead women and babies at Wounded Knee. That's why we have Song My and My Lai now.

To us life, all life, is sacred. The state of South Dakota has pest-control officers. They go up in a plane and shoot coyotes from the air. They keep track of their kills, put them down in their little books. The stockmen and sheepowners pay them. Coyotes eat mostly rodents, field mice and such. Only once in a while will they go after a stray lamb. They are our natural garbage men cleaning up the rotten and stinking things. They make good pets if you give them a chance. But their living could lose some men a few cents, and so the coyotes are killed from the air. They were here before the sheep, but they are in the way; you can't make a profit out of them. More and more animals are dying out. The animals which the Great Spirit put here, they must go. The man-made animals are allowed to stay— at least until they are shipped out to be butchered. That terrible arrogance of the white man, making himself something more than God, more than nature, saying, "This animal must go, it brings no income, the space it occupies can be used in a better way. **The only good coyote is a dead coyote.**" They are treating coyotes almost as badly as they used to treat Indians.

Some people's uncaring attitudes toward life forms and other people is but one of the many detrimental side effects of prejudice against Nature.

By living with Nature, I have discovered that I formerly accepted the dismemberment of creatures, canyons, and cultures, because through my addiction to technological euphoria, I felt personally dismembered. Fifteen years ago, I subconsciously recognized that in order to enjoy the fruits of my culture, I had been amputated from Nature. I felt cut off, and my subconscious told me that Nature was the culprit, and conversely, power, education, control, technology, and religion were my saviors. To avoid further discomfort, I confusedly attacked the alleged culprits, the plant and animal kingdoms, under the guise of prosperity and self-preservation. I set myself up in business for the money that was in it, money that I felt was needed to increase my distance from Nature in the form of cars, television, fashionable clothing, a large apartment, and other material things. I was filling my euphoria deficiency needs. Unbeknownst to myself, my lifestyle had a high impact on Nature.

Indoctrinated from birth that our natural feelings and drives are suspect, most of us learn to despise our *uncivilized* natural selves. We strive to cover our blueness. We are taught to crave the red-cubic human-constructed environment and the social graces of our mass-culture society. We invent gods and pacifiers to tranquilize the human-created anxieties that grow out of our Nature-alienated existence. Our anxieties are often exploited; but seldom are we given the opportunity to recognize their source, or to reduce them permanently.

If a human infant were placed in a closet, away from normal contact with people, and were given what it needed to survive, what kind of person would emerge in forty years? Would that individual be considered human? What human qualities would it have? The growth of nervous tissue and connections within a person is entirely dependent upon the quantity and quality of that person's experience with the total environment because, psychologically, an infant and the environment into which it grows are identical. Nerve growth would be severely stunted in a person kept in a closet, and the adaptability of that individual would not approach its full human potential. The closet would have blocked off vast energy exchanges, relationships that would have ordinarily taken place between the infant and life in the outside world.

A society that is prejudiced against Nature is a red-cubic closet with respect to the infant's relationship with Nature. Every aspect of Nature thrives within the human child. The ecosystem beckons the baby from the womb of its human mother to its new womb in Mother Nature, the planet Earth. The child hears the callings of Nature through its Natural feelings such as thirst, hunger, excretion, loneliness. These feelings and needs are the voice of Nature; they are met and satisfied by Nature's global gifts and love.

Because much of our society is prejudiced against Nature, it closets off the infant's relationship with Nature. The baby's nervous system is thus thwarted from reaching its full natural potential. The natural drives and needs of the child eventually strike our restraining, anti-Nature taboos like raw eggs hurled against a brick wall—splattering social and environmental ills over everything in sight. The red-cubic child develops to adulthood knowing full well the vise-like strength of the brick wall of its culture.

Sometimes we incorrectly identify the brick wall of restraining taboos as God.

Which individual should we emulate to relieve us of our social and environmental problems, we ask ourselves. Who will show us the way? Perhaps the answer is God. Jesus, Jehovah, Mohammed, Buddha, Wakan Tanka, Krishna, Taiowa, The Great Spirit: they all have the same unattainable worldly powers and knowledge. Each god is fully related to the cosmos; each is in equilibrium with the universe; each asks the same of us in His own special way. It is people that are prejudiced against Nature, not our gods. Our gods were found and are revered to help us deal with the miserable results of our prejudice against Nature.

Living with the Expedition community in the changing environments we encounter across the face of America teaches us that God is not a brick wall. He/She is the spark of life that encourages a person to grow harmoniously with Nature and each other.

What kind of people do our gods tell us to be? I can't speak for your

God. But the past decade has taught me to reach for the exhaltation, the euphoria, of harmony with God, Nature, and humanity. I desire a stable, supportive family association of man, woman, and child, and to develop close cooperation between my family and other groups of people. I desire to be wanted and useful, to avoid loneliness and isolation through beneficial interdependence. I continue to seek educational opportunities that fulfillingly adapt and prepare me for a positive role in society. I desire a gratifying sense of work from the personal contribution of my daily energies to society. I desire the security of a sense of place and community. I desire a marriage held together by love, survival, mutual respect, and cooperation. I seek the leisure to enjoy the gentle land, the sparkling waters, the interesting life forms, and the friendly people that make up my environment. I desire to live in harmony with the wealth of beauty and life found in my countryside, and with the excitement and wonder of the life process. That is the command of my God. Is it not the command of all gods, no matter who and how we call them? Is it not euphoric?

It is a slanderous contradiction that those of us who desire such utopian relationships in our personal lives on Earth are called unrealistic dreamers and idealists. The eye-opening truth is that the only people ever to come consistently close to attaining the above-described utopian social and emotional ideal are the people whose culture and life did not harbor undue prejudice against Nature. They were hunters and gatherers, or close to it. The Amish, organic farmers, and alternative intentional communities, to some extent follow suit.

To the hunter-gatherers, Mother Earth was a continuation of the relationships that were established within the human womb. They revered the Earth. They enjoyed the quality and goodness of life that we identify as a mandate from our gods, that we increasingly destroy as we seek it in vain due to our prejudice against Nature. One has only to meet or read about hunting and gathering people who live close to the land to know the equilibrium of their lifestyle, and to personally desire its values. They were and are people who lived and loved their global blueness. The scriptures seem to describe them: *He has made everything beautiful in his time; also He has set the world in their heart.*

Our history reveals that Americans learned very little from the demise of the Garden of Eden and from the crucifixion of Jesus. We repeated these acts 1500 years later with the conquest of North America and the destruction of the hunter-gatherer peoples here. It might well be that Jesus Christ returned to our lives, and, unbeknownst to us, died in a museum in California. He came not in the sterile, golden glory of exalting technological euphoria, but rather in the simplicity and joy of a man called Ishi, the last native hunter-gatherer known in our country.*

Lame Deer:

    The buffalo is very sacred to us. You can't understand about nature, about the feeling we have toward it, unless you understand how close we were to the buffalo. That animal was almost like a part of ourselves, part of our souls. The buffalo gave us everything we needed. Without it we were nothing. Our tipis were made of his skin. His hide was our bed, our blanket, our winter coat. It was our drum, throbbing through the night, alive, holy. Out of his skin we made our water bags. His flesh strengthened us, became flesh of our flesh. Not the smallest part was wasted. His stomach, a red-hot stone dropped into it, became our soup kettle. His horns were our spoons, the bones our knives, our women's awls and needles. Out of his sinews we made our bowstrings and thread. His ribs were fashioned into sleds for our children, his hoofs became rattles. His mighty skull, with the pipe leaning against it, was our sacred altar. The name of the greatest of all Sioux was Tatanka Iyotake—Sitting Bull. When you killed off the buffalo, you also killed the Indian—the real, natural, **wild** Indian.

Prejudice against Nature has blinded many Americans as it once blinded me. We cannot perceive that the human womb is but a starting memory of how it feels to live harmoniously in the womb that is the planet Earth.

America's denial of Nature has left the egg of our thwarted drives splattered all over the pages of history and today's newspapers. It instigates war, hatred, tension, violence, loneliness, extinction, pollution, competition, drug use, alcoholism, mental illness, and immorality. None of these aberrant behaviors were typical of the Shoshone societies of hunter-gatherers. They were sensitive to Nature's fluctuating but harmonious life-giving moods, worshiped them, and regulated their lives around them. We stole a continent that had been a Garden of Eden for time immemorial.

---

*See *Ishi in Two Worlds* by THEODORA KROEBER, published by the University of California Press, Berkeley, California.

After less than four hundred years, will we leave it for future generations as a garbage pit of pollution and social stress?

A recent Expedition Institute visit with the Old Order Mennonites and Amish (the horse-and-buggy Plain People of Lancaster County, Pennsylvania), dramatically illustrated to us the congruity that could be obtained by applying the Whole Life factor to modern life. Because their culture preserves the values of 16th century rural life in Europe, it also contains some whole-life relationships between people and the land, relationships that have not been extensively interrupted by dependence on and overuse of modern technological assistance.

With the Plain People, it is religiously taboo to be extravagant or to oppose church counseling in any aspect of material, social, personal, or religious life. Church counseling by parish members' consensus results in a stabilized "distance from Nature" increment which does not place increasing demands on the land. Profits from farming are not needed to purchase excessive goods or services.

By using low-impact horse-drawn farming techniques that incorporate much family labor and attention, the Plain People keep their identity by remaining independent of modern mass-culture technologies, fashions, and practices. This reduces their economic pressure on the land, because theirs is a slow "inefficient" farming method in comparison to the use of quick energy-intensive high-profit, powerful tractors and equipment. It gives the land time to regenerate.

Unpaid, long hours of jack-of-all-trades family farm life on small acreages are the Plain People's way of life. Pleasures are obtained from loving rural living, its challenges, and strong family, community, and religious ties. A vital, institutionalized commitment and feeling for God, people, and moderation tempers their profit motive and provides the gratifications of independence, self-reliance, and self-sufficiency. This relieves the land of the pressures necessary to sustain a profit in order to gain these same gratifications by sustaining distance from Nature with material wealth and excessive mass-culture technologies.

Cooperation is the key to the Plain People's feeling of community and place. Insurance policies for health, property, fire, and crops is unknown. When problems occur, neighbors pitch in and help to maintain the unfortunate victims of disaster. The costs of insurance and social security are eliminated, further relieving pressures for profit from the land.

Most of the Plain People's food is wholesome, homegrown, and inexpensive. Our host's family of six persons spends a total of $2,500 per year to completely maintain themselves (1982). They paid off a $75,000 farm mortgage in seven years. They claim that they were hardly affected by the depression. Their children are experientially educated to be independent family farmers just like their parents and neighbors; they are pacifists and not subject to the draft.

A traditional, sensible, and concerned land-use crop rotation system is practiced by the Plain People. Although it would be unprofitable in our mass culture, they thrive on it. So does the land. It has been farmed for 250 years and continues *to increase its production of topsoil* in good growing areas, while the alarming national trend is toward topsoil depletion.

The relative health of a farm ecosystem can be an accurate indicator of the health of its inhabitants' relationship with Nature and each other. The land reflects the consciousness of the culture that inhabits it.

Although the Plain People achieve the desirable holistic end result that might be obtained by our society's applying the Whole Life factor to farming today, they are unique in that they are the continuation of a way of life that evolved in a different era. Each of them is born into a strict subculture that maintains its identity by abstaining from participation in our wasteful, mainstream way of life.

We have also visited with modern Americans who have purposely adopted holistic farming principles. They apply the Whole Life factor by practicing organic agriculture. Many of them have indicated that they can happily sustain their low-impact lifestyles as well as their farm and customers. They attempt to:

- Maintain and improve soil fertility as well as the purity of water and air by applying newly-devised recycling techniques, as well as those extant in Nature but too frequently disrupted in our time.

- Apply recyclable materials to the soil in such a manner as to avoid all forms of pollution.

- Produce food of the highest nutritional value in sufficient quantity to maintain people's health.

- Convert to renewable energy resources and minimize the use of fossil energy in agriculture.

- Furnish for all livestock an environment that provides for their physiological needs and conforms to humanitarian principles.

- Make it possible for agricultural producers to earn a living through their work and develop their potentialities as human beings.

- Encourage environmental protection and a respect for the living matter with which the agriculturist works as an ally.

- Support public policy that furthers these objectives.

Actualizing the Whole Life factor is effective in approaching congruency with the planet, but it goes against the current of mainstream America, because the latter is prejudiced against Nature. Our mass culture needs more energy from the planet in order to excessively increase people's distance from Nature. As a comparison, in Haiti the average person's energy consumption is equivalent to that produced by burning 68 pounds of coal per year, while the energy consumption of the average American is 23,000 pounds of coal per year and is expected to double by the year 2000 A.D. We are goaded into consuming.

Before I became aware of the effects of my lifestyle upon the continent, I was subject to the messages of the media. Our desire to increase our distance from Nature is intensifying as the impact of the media is increasingly felt. Many Americans are becoming more bewildered. Many of us have yet to recognize and cope with our prejudice against Nature and its consequences. We continue to emphasize the importance of human interrelationships and ignore our lack of close relationships with Nature. It is our greatest error, for the latter stymies the former.

At this late date, it seems that most people would choose the environmental impact of a power dam or the extinction of a species, rather than give up the technology, status, and distance from Nature of an air conditioner. Many of us have been emotionally dismembered, and we are unable to remedy our dismemberment because we are prejudiced against the planet that could heal us. We are crippled by our addiction to excessive technology.

With the aid of technology and the power to symbolize, people can be as gods. We need only to believe that a canyon could be a lake, and it may become one. Our conceptions can move mountains, create cities, or annihilate communities. One would think it would make little difference if half a glass of water were perceived as half-full or half-empty—but such subtleties in the minds of gods can inflict severe consequences upon the process of life.

Lame Deer:

The green frog skin—that's what I call a dollar bill—that was what the fight at Little Bighorn was all about. The gold of the Black Hills, the gold in every clump of grass. Each day you can see ranch hands (including Indians—*MJC*) riding over this land. They have a bagful of grain hanging from their saddle horns, and whenever they see a prairie dog hole they toss a handful of oats in it, like a kind of little old lady feeding the pigeons in one of your city parks. Only the oats for the

prairie dogs are poisoned, because they eat grass. A
thousand of them eat up as much grass in a year as a
cow. So if the rancher can kill that many prairie dogs
he can run one more head of cattle, make a little more
money. When he looks at a prairie dog he sees only a
green frog skin getting away from him.

For the white man each blade of grass or spring of
water has a price tag on it. And that is the trouble,
because look what happens. The bobcats and coyotes
which used to feed on prairie dogs now have to go after
a stray lamb or a crippled calf. The rancher calls the
pest-control officer to kill these animals. This man
shoots some rabbits and puts them out as bait with a
piece of wood stuck in them. That stick has an explosive charge which shoots some cyanide into the
mouth of the coyote who tugs at it. The officer has
been trained to be careful. He puts a printed warning
on each stick reading, "Danger, Explosive, Poison!"
The trouble is that our dogs can't read and some of our
children can't either.

And the prairie becomes a thing without life—no
more prairie dogs, no more badgers, foxes, coyotes.
The big birds of prey used to feed on prairie dogs, too.
So you hardly see an eagle these days.

Like other social ills, competition is a side effect of prejudice against
Nature. It underlies most of our relationships even though we are aware
that it is a questionable motivating force. Competitive feelings overwhelm
the more subtle values to be found in life relationships and activity, yet
competition is a long accepted and documented part of Nature. From
Darwin to any modern ecologist, competition is the solidly based explanation of why some plants, animals, or people are able to survive in a specific
habitat. The key to survival, evolution, and ecosystems is that the best
competitor for water, nutrients, sun, or air is the one which survives. Why
shouldn't competition be an accepted part of human behavior? People are
natural animals. People feel competition. It is part of us. Why deny it or
disapprove of it?

Let's consider the role of competition in humanity. Is it real, or is it a
destructive manifestation of prejudice against Nature?

At birth, the helpless infant leaves the euphoric atmosphere of chemical support and physical protection of the womb. It encounters the outside
world process of life, its womb securities buried in its unconscious

memory. The child learns to compete for technology and approval which it experiences as a replacement of the security of the womb. The growing child vies for money and the imagined securities money can buy: food, shelter, protection, approval, chemicals, and euphoria. The child is encouraged to become a competitor, a winner. We are *born to win*— everybody loves a winner. Why should this competitive process be questioned?

Consider the effects of competition in both humanity and Nature. In humanity, the effects are detrimental both to interpersonal relationships and to the ecosystem. Encouraging competition leads to perceiving the globe as a resource to be competitively exploited for its security and life-giving commodities. This is a dangerous perception; the planet Earth is actually a whole, delicate, intricately balanced, interdependent energy system and life process. It is *not* a resource; it is *our* source.

If we experience the process of life to be based on competition, we encourage competition. The result: people relate to other people, places, and life forms, competitively. Competition leads to exploitation, isolation, and strife, the antithesis of conservation and peace. Because of these negative effects, competition should be rejected as a reasonable way for people to relate to the planet or to each other.

Although it is detrimental in human populations, competition is the basis of adaptation in Nature and is considered to be a positive, necessary process. Are humanity and Nature different from each other? No. Humanity is part of Nature, and that leads us to an otherwise unthinkable conclusion! We have been deceived about the nature of Nature.

Humanity's historic prejudice against Nature has colored the observations of Nature made by leading scientists and philosophers throughout history.

Consider competition as it exists in Nature. In Nature, we find that competition between organisms, and between species, stems from two basic relationships that are counterparts of each other:

1. Each organism and species produces many more offspring than are needed to replace themselves; these progeny compete for habitats, niches, nutrients, and survival.
2. The planet has a limited amount of natural resources, and all organisms must compete for these limited resources in order to survive.

What is phenomenal about such observations is that they lead even the most naive observer to conclude that the removal of some organisms from a population would place less competitive pressure upon those that remain. Life would, therefore, tend to flourish and be stronger for their removal. Yet, actual observations prove the opposite to be true. In the long run, the removal of organisms from the ecosystem breaks the web of life and weakens the system. How can this be explained?

Let's build theoretical models of Nature which meet the observed facts. In one model we arbitrarily eliminate the use of the concept of competition. Not only does that model work, but it is congruent with our observations and experiences in human communities.

The key factor in the design of an acceptable model is that, in it, we replace competition with its antithesis, cooperation. Our reasoning is based on three lines of thought which we are forced to use because we have excluded the cultural concept of competition:

> 1. The overproduction of offspring in living populations produces *stress* on each individual and species.
> 2. The organisms or species that best *cope with stress* are the organisms that survive.
> 3. The organisms or species that, in time, best establish cooperative interrelationships with the environment are most adaptable to stress and are, therefore, most likely to succeed in any given habitat or life zone.

I must emphasize: fitness and adaptability are *based on the ability to cooperate,* not compete.

Competition stems from ignoring the value of feelings like stress, and from human anxiety about the availability of raw materials for *over* protection from Nature.

These very same observations were made of the Shoshone hunter-gatherers. According to Peter Farb, they did not compete, nor know war, because "they had no territories to defend, for a territory is valuable only at those times when it is producing food, and those were precisely the times when the Shoshone cooperated, rather than made war."[1]

A society is a means of coping with the normal T-R stress of survival in Nature. In time, a society's survival goals can successfully be met through interpersonal and environmental cooperation. The Amish, the Hopi, some organic farming communities, and populations of social animals such as wolves, exemplify this cooperative society. Our survival efforts should go into developing cooperative relationships.

A cooperative conceptualization of Nature is quite different from the fearful, aggressive picture typically painted of relationships in Nature. Nevertheless, aggression and territorial behavior are important functions of all natural cooperative relationships. Aggression and territoriality serve as a *language* which delineates and defines the existence of each organism and its immediate support community, thus limiting the amount of stress and strife placed on the interrelationships within each niche.

Historically, in America and elsewhere, the inability of native biotic and human populations to *cooperate* with foreign invaders—and vice versa—has led to the demise not only of native populations, but of the invaders as well. We have incorrectly believed that we are exempt from Nature's system of checks and balances, that we could, without harm, competitively change whole environments and nations as we psychologically and physically strive to avoid Nature. The T-R web of life is integrated and balanced. It is pulled by stress and held together by cooperation. We have inaccurately imposed our unempathetic competitiveness for resources upon the cooperation of the universe and are the poorer for our transgressions.

Whenever the concept *competition between organisms* (and that includes people) appears in discussions or in literature, the reader may acceptably and more accurately substitute *long term cooperation*. The substitution correctly relegates competition to its status as a figment of humanity's anxious imaginations about Nature and each other.

As gods powered by technology, it is beneficial for us to substitute cooperation for competition with Nature. Our perceptions and actions must become congruent with holistic global realities, if we are to approach equilibrium and avoid harmful side effects. Cooperation with Nature provides mutually rewarding, immediate and long-term relationships. It stops our dismemberment and encourages disarmament. It returns us to a source of survival that was lost when the eggs hit the wall.

§

Contact with the Expedition environment and the kaleidoscope of North America has demonstrated to me that every environment tries to dictate what we are at any given moment and then proceeds to condition us to remain that way—especially when we are young. If we desire to become whole, cooperative people, who are not prejudiced against Nature, then we must create an environment that conditions us appropriately, an environment for and in ourselves that is whole, operates cooperatively, and teaches how to gain good feelings by developing harmony with Nature's fluctuations. This is the quest of the Expedition Institute as it uses the American environment and the Expedition community as a classroom and teacher.

The trouble is that the already established technology
and present-day culture into which we are born is an

immense authority. It not only *turns on* childhood discomfort, but it rarely confronts us with the questions,
with the problems. It merely supplies us, as children, with
the already established answers, while we are immature,
and we are brought up to live in and accept them. "You
live in a house, not a hogan." There are few or no questions to be asked in areas where, during early childhood,
answers have already not only been provided, but
accepted as well. Educationally, this is unfortunate, for
again, without real questions, answers seem to have little
meaning.

Because we don't really question much of modern life
and didn't arrive at it ourselves as children, for many of us
as adults it has little real meaning; and we treasure only
what does have meaning in our daily existence, mainly
money and the technological assistance that can be
bought with it. In addition, we teach our children what is
important to us, what we treasure, and so another generation appears with the same values as the previous one,
and also with the same, or seemingly more intensified
problems. Here again, we see the straight-line syndrome
in action, going around in circles and perpetuating itself
and its problems as well. Is that who we are? It probably
is. Why not?

To break the circle, individuals must first be aware of
the existence of the circle and of their feelings as well.
They must then learn to start asking questions about
both—questions that have meaning to them. They must
begin to examine evidence so that the answers they have
personally discovered will have a basis in reality rather
than only in childhood feelings and imagination.

Educationally speaking, how do you get individuals to
start asking questions? My thought is that a most
productive way would be *to remove them from the
environment that prevents them from finding questions,*
from the overprotected, technologic, indoor home and
school environment of their childhood, which, for the
most part, provides only the answers. To this end, we have
created an educational expedition dedicated to discovery,
which, it seems to me, is a strong step to returning to our
origins and reconsidering how we will live and relate in

the present and the future. The expedition unquestionably achieves this purpose as it solves the problems that
prevent it from doing so.[2]

The Audubon Expedition could be accurately considered to be a 1-4
year accredited *ropes course* with the destructive relationships between
people and places in America as the problems to be solved by applying the
Whole Life factor to them.

An outsider's view of our program in action is shared by a reporter
who spent three days with our learning community for the *Tucson Daily
Citizen:* (For more information, see Appendix C.)

§

Maybe we've been going about this education thing all wrong.

Maybe the answer isn't more classrooms or razzle-dazzle technology—a
computer terminal in every home, for example.

Maybe, instead, it's something very much like the National Audubon
Society's Expedition Institute, a schoolhouse on wheels that has stopped
off in Tucson for about 10 days.

The Institute doesn't look much like a school. In fact, it looks more like a
backpackers' campsite, much too casual and too much fun for the often
painful process of stuffing knowledge into the brain.

The whole thing becomes suspect when the 20 students, ranging from
some in high school to others at the master's degree level, rave about how
much fun they're having.

But it's hard to argue with the results. Many of the students go on to
college after completing their junior year of high school, and the Institute
and graduates of the program say they're far ahead of their peers when they
return to more traditional settings.

One of the students explained, "If we want to study the desert, we don't
read a book about the desert; we go to the desert."

The school, which consists of a large yellow school bus crammed with
books, food, musical instruments, camping gear and other equipment, goes
to places such as the Florida Everglades, prehistoric Indian sites, Amish
communities in Pennsylvania and the Environmental Protection Agency in
Washington, D.C.

Mike Cohen, one of the three adult guides at the school and the man who,
with his wife, Diana, organized the 13-year-old program, said the students
learn to question the natural environment, their social environment and
themselves during the September-to-May program.

"They answer their own questions, not someone else's questions," he
said. "About 90 percent of what's taught in traditional schools is forgotten
after a year," Cohen contends, "because it wasn't learned out of curiosity,

but for a grade."

If they want to learn about Mexico's Tarahumara Indians, for instance, they don't read what someone else thinks about the Indians (although that might be done later). They go and talk to the Indians themselves.

In fact, they have just met Tarahumara and Papago Indians at a loosely organized music and dance party sponsored by Tucson anthropologist Big Jim Griffith.

The students currently are camped at Griffith's place just outside the Papago reservation near San Xavier mission.

Griffith is one of about 150 experts, specialists and resource people the school draws upon around the country.

Every 10 days, taking advantage of the compact library on the bus as well as university libraries around the country, each student produces a paper. Then everyone else reads all the other papers and they're discussed. Students react to the grammar, style and content.

All this is the academic aspect of the school. But there's more to education than mental calisthenics. While the intellect gets a good workout, other facets of the program exercise the students' judgment.

They learn to shop for themselves, feeding 23 people on $80 a day. Chores are handled by work committees, which are reorganized periodically so each student gets a chance to work with everyone else in the school. And emotional maturity comes in working out differences among themselves and meeting as a committee each day to decide what will be studied the following day.

The students said they avoid *cabin fever* while living in close contact with one another by bringing feelings into the open and discussing them. In the process, they become very comfortable with one another and self-confident. They form a closely knit community, but one in which a stranger is immediately made to feel at home.

By the standards of 20th-century urban America, the school is austere. In fact, by the standards of 19th-century America, the school is austere.

As in Thoreau's camp near Walden Pond, material life is reduced to basics. There is no TV, no radio, no stereo; just the tents they camp in each night, the battery-powered lights inside the bus and the Coleman stoves they cook on.

By all American standards, these people are deprived. They don't even have access to a shower every day.

They have more to do than they have time for, however, unlike many teenagers

Frank Trocco, another adult guide, said that's probably because most teenagers aren't given any responsibility for their own destinies.

In the Institute, however, the students have a total responsibility for themselves. There are no authority figures. Trocco said he and Mike and Diana Cohen act only as catalysts. All the organization for the program is in the hands of the students.

"Every evening they deal with problems that come up," he said. "They get the solution plus the process of dealing with the problem that very few teenagers have contact with."

The group also has the strong awareness of the environment that comes from living in a non-electric kind of way. Without TV, movies or recorded music, students must make their own entertainment. Several have learned to play musical instruments. Two play the fiddle, three are on penny whistle, two play guitar and one picks banjo. Everyone sings and dances.

Other forms of entertainment include hiking and camping, which are educational experiences in themselves; one student said: "It's a way of enjoying things that isn't destructive to the environment."

They are proud of their low-impact lifestyle. By a consensus decision this year, they have limited themselves to one paper towel a day per person and there are no Kleenex boxes around. Everyone carries a handkerchief. If they didn't, the paper waste would be phenomenal, they said.

The Audubon Society, traditionally an innovator in environmental education, began sponsoring the Institute because it concluded that many conservation problems are caused by the way young Americans are taught to live and relate to the natural world. The program operated on its own under the Cohens' direction for nine years before Audubon sponsored it.

"This institute can be done on a much larger basis," Trocco says. "And that's what Audubon plans to do, " he added.

The institute is starting an intern program to train leaders to take groups like this one around the country. Information about the intern and regular programs can be obtained by writing to: Audubon Expedition Institute, 950 Third Ave., New York, NY 10022.

"If you get trained leaders who can guide the students, it will work," Trocco said. "After all, all these young people want to do is grow."[3]

REFERENCES:
[1]FARB, PETER. *Man's Rise to Civilization,* p.53. New York: E.P. Dutton, 1968.

[2]COHEN, MICHAEL J.*Our Classroom Is Wild America.* Cobblesmith, 1978.

[3]STILES, EDWARD. Audubon Society 'Schoolhouse': the un-technological revolution. *Tucson Citizen,* March 19, 1979.

Summary and Conclusions

There are many detrimental side effects from prejudice against Nature: uncaring attitudes towards life systems including people, subdividing life systems which results in their demise, creating and stressing competition rather than cooperation as a survival mechanism, alienating ourselves from our internal, inborn nature, and depending upon superpower Gods to deal with the problems that emanate from the folly of our poor relationship with the planet. The less prejudicial societies of the hunter-gatherer people, the Amish and organic farmers act as controls against which our quality of life and survival potential can be measured. Their environments are better able than ours to support life. Expedition Education attempts to reduce our biased acculturation's side effects by giving young people the opportunity to remove themselves from environments that reinforce their prejudices against Nature, and to develop the skills for establishing more Nature-compatible life relationships.

Chapter Ten

WORDS WITHOUT KNOWLEDGE

Mother Nature is the universal womb of life. Her life-giving relationships are duplicated in the womb of our human mothers.

There is a new *Gulliver's Travels* story that bears repeating. Gulliver was shipwrecked and washed up on a remote island in the Pacific. He was amazed to discover there that every member of the flourishing population was blind from birth. On the island, Gulliver was befriended by a very accommodating gentleman, Toby, who taught Gulliver the native language and introduced him to the island's customs.

As Gulliver settled down to live in the blind community, he began to make what he felt to be simple, reasonable requests for materials that would help him feel at home. With the exception of Toby, who had become accustomed to his peculiarities, the island people were shocked by Gulliver's requests. They were sure he was insane—possibly dangerous. It was their natural reaction to Gulliver because they were blind. Never having been exposed to a person who could see, they could not understand his arguments in favor of books, pens and paper, glasses, signs, a bicycle, and a camera. His continued requests for these items and others were experienced as personal affronts by the blind citizens, because they were outside each individual's configuration of consciousness. Gulliver unwittingly made them feel stupid and threatened, as he persisted in his efforts to explain his desires clearly so that people would get to know and understand him.

Finally, Gulliver's presence on the island became so irritating that the community was set on jailing him. Only Toby came to his rescue. He had Gulliver examined by the island's leading medical team of scientific researchers and practitioners. (All were blind, of course, but extremely capable.) After several days of testing, these most brilliant of doctors discovered that Gulliver had an extraordinary condition which was not, in fact, insanity. He had a bad case of a rare disease known as *eyesight*, and it was determined that the disease was caused by his unusual growths called eyes. The medical team claimed that they could cure Gulliver's malady with a complicated operation that would remove his eyes. Toby was overjoyed at the news. "Thank God for science!" he exclaimed.

Wouldn't it be fascinating if a modern Gulliver visited America today? Like the Gulliver of the story, the Gulliver I have in mind would be able to see in areas where we are blind. Our Gulliver's perceptive gift would be that he was still able to see Planet Blue; to see how life survived; to see how the planet maintained itself, and therefore us. Unpainted by our cultural conditioning, he would see the planet Earth to be a source of human existence, and that it surely must itself be alive. He would easily perceive that our civilization has walled us off from Nature's communication, has prejudiced us against Nature, and has spawned crippling environmental and social problems. He would be able to see our anti-Nature prejudice, because he would not have been brought up *blind*. He would not have been educated, as we have been, to justify our prejudice against Nature as progress, academics, economics, or standard of living.

It is questionable whether Gulliver's ability to see Planet Blue would be considered a gift in today's society. Because of it, he would probably meet severe rejection.

Such was the case with the contents of this book and Susan Gould. The book's paradigm lies outside of the perceptual parameters of mainstream America, because we have not been sufficiently exposed to an environment that contains enough substantiation of the paradigm.

It's only if and when an individual purposely seeks evidence and experiences that lend substance to my suggestion that we are prejudiced against Nature that the validity of that claim begins to appear. It is an exercise that the readers may choose for themselves, but mine was a different path of the same end. I was never given the choice. Instead, I was unknowingly exposed to the evidence from Nature and my feelings.

For fifteen years, I have lived with Nature, gradually increasing my ability to hear her voice and to understand her message. I am not saying that I have visions in the night or auditory hallucinations of voices from afar. What I am saying is that Nature speaks a language that we are seldom taught to understand, and thus, few can hear. Look in the language section of any college catalogue; you will find Latin, Spanish, French, and German, but you won't find feeling. T-R feelings are Nature's native tongue. Feeling creates and communicates relationships that are based on trust and understanding rather than on contracts and money.

One way we help expedition students sense Nature's voice is to suggest they write a subjective description of an environment that they choose to observe. Although difficult to create, the resulting essay is unique because it contains no objective thoughts. Instead it communicates reactions and sensations that originate from prolonged contact with the landscape. After a while, most people who attempt such an essay discover that they can better "hear" the environment; they produce interesting papers that convey Nature as a feeling. The exercise improves student's writing as do their subjective letters of thanks to the resource people who share their lives with us during the year.

I first became aware of the power of feelingful perceptions of my surroundings eight years ago, when the Expedition was exploring Big Bend National Park in Texas. It was our first exposure to the West that year, to deserts that had been created by the cattle industry, and to the sight to mountains that were not covered with trees.

I was trying to describe the size of the Chisos Mountains in an essay. (Yes, the staff participates in group writing assignments.) I found that I could not communicate their size, because I had nothing with which to compare them. There was no control or measuring stick present. I had memories of the Tetons, the Sierras, the Green Mountains, the White Mountains, the Olympics, the Adirondacks, the Smokies, and the Cascades, but images were not enough; the ranges themselves were not present for comparison. I was helpless to describe the size of the Chisos without using unavailable numerical and comparative measurements.

After several attempts at my essay, I realized that my preoccupation with exact size and comparison was a manifestation of my culture. The Chisos were whole; they were a life system; they were the Chisos. To describe them, I could use only my direct observations and my feelings about approaching them. I believe that this is the way the natural mind functions without cultural training. It soon dawned on me that these perceptions and emotions were in their own way comparisons; I was comparing the Chisos to myself. I was relating directly to the mountains, and they were influencing me in their turn.

Similarly, I once found a small, lovely piece of driftwood at the bottom of Escalante Canyon. I kept it, and have used it to prop open a window in my office for the past seven years. If I were asked to describe that piece of driftwood now, and were to compare my present description of it to my perceptions of it seven years ago, the portrayals would be quite different. Time, contact, changes in the way I think, and the history of my relationship with the wood over seven years, all would influence my description of it now. Or it could also be said that the driftwood itself had influenced me and that I, therefore, began to know it differently. Was the wholeness of the driftwood any different than the wholeness of the Chisos Mountains or of myself? Did, in fact, the wood speak to me—reach me—in its own way?

We experience a house differently after having lived in it for seven years. Can't we truthfully say that the house has spoken to us, so that we know it better than when we first moved into it? If we desire to accurately consider cause and effect, it seems fair to say that our wholeness can, and does, relate to the wholeness of both inanimate and animate living things. And vice versa, it seems fair to say that they can speak to us.

I have lived with Nature for nearly two decades. I leave it to your judgment whether I am justified in saying that Nature speaks to me and describes to me how we are prejudiced against her. In my opinion, it is no more insane for me to believe that Nature speaks to me than it is for anyone to believe that he or she loves his or her car or home or country. Both are simply different perceptions of the same emotional phenomenon.

§

Mother Nature has told me that she is concerned about her family relationships, how her children are getting along with each other. She is worried about our self-preservation. Like an insurance company, she wants each of us to stay alive as long as possible, because as part of her we all support her, and each other as well.

We're a nice enough family, except that, of all her plant and animal children, humans, especially Americans, give her trouble.

If you can put up with Nature's sentiments, you can understand her feelings. She's generally quite good-natured, but she is very upset these days to see how we're prejudiced against her.

I hear her wise, weakened voice saying:

"Mike, it's not easy for me to speak to your civilization. I'm from the old country of trees, plants, and mountain-to-mountain buffalo. I'd never even heard of the *three R's* until I heard some Greek declaiming about them. So I've had my difficulties with all my human children.

"But, Mike, you American kids were always different from the rest. I've never been able to talk much to you, and I've never really known why. I've tried, from the minute you got over here, to lend you a helping hand so your lives would be happy—but not many of you paid particular attention.

"I talked to *your* feelings, Mike, for eight years before you could begin to understand me. You still don't catch everything, but you're improving—and I have to tell my children what's on my mind.

"I've studied Americans, and I think I understand now why you don't listen to me. You ignore me because I'm your mother. *Mother.* You human children gave me that name just as soon as you invented speech. Back then, you weren't so confused. Back then, you could plainly see the real relationship between us. We were close.

"But now, kids, now you're set on treating me like most adolescents in your culture treat their parents. What happened? Where did I go wrong? You perceive me to be interfering and unfair, an infringement on your rights of freedom and independence. You know-it-alls act like you can create and maintain your lives.

Obviously, I make you feel helpless, as though you don't know anything. That makes me unhappy; I'm your *mother.* But somehow I feel you wouldn't have anything at all to do with me, if I didn't still give you your allowance, and pay, pay plenty for your education.

"Mike, you have to let them know what I'm telling you. You have to make it clear so they can understand—so they won't discriminate against me.

"As long as you're alive, you will always come to me for your allowance. Warm sun, clean air, clear water, healthy food, and strong bodies—that's our relationship. I hear you when you say you want to be independent, but...I'm your mother; you know I want only what's best for you, and I have to tell you: you can't live without your allowance. I'm not talking this way to be hard or cruel; it's the simple truth. Without that allowance, you can't exist. Do listen! I'm not trying to be unreasonable. I love you; you're part of me. I'll give you your allowance, rain or shine, as long as I possibly can. I'm not threatening you. All I'm asking is that you have a good, mutual relationship with me, that at least you feel good about me, respect me, and treat me kindly. It's hard to understand why you refuse to do that.

"Ever since you got to America, you've been set on destroying me. That sounds ugly, doesn't it? Nevertheless, you've been trying to kill me, your mother. Where did I go wrong? You've tried to dissect the way I work, so that you can more easily grind up my lovely body and manufacture a little short-term euphoria. I can see that you want to avoid the normal fluctuations of hot to cold, wet to dry, but you've gone way past that point. You've gone into business, making pocket money out of my life and organs. Money! You believe it will give you full independence from me, for in spending it you're able to make your own short-term euphoria whenever you want it. It's so important to you that you've made it a federal offense to deface a dollar bill—while it's perfectly okay to carve up a mountain! You're money-made euphoria doesn't work in the long run. I'm sorry if telling you so makes you angry, but there it is.

"I, Mother Nature, am the universal womb from which you came and in which you presently live. I am the womb from which your comfort is naturally derived. I am the source of humanity, the source of the human womb that provided those euphoric feelings, when you were but an embryo with a memory for feelings. I am those euphoric feelings. I'm concerned—both for you and for me. To get back to basics, where in the world would your children find euphoria, if you killed me, if I were dead and gone? How would you live?

"Already your children are forming in human wombs that are polluted with the noxious poisons you breathe, eat, and drink. Today, the hormones produced by living without my close support are affecting your children. They seem to have less of a desire to live and be healthy than had the children of just a few generations back. And even those who do desire a wholesome life are having a hard time achieving it, because the equilibrium of the planet has been so altered by all your activities. Now there's hardly one of my streams from which you can safely drink; the very air you breathe is bringing poisons into your bodies.

"Why don't you listen to me? You can still hear my voice in your feelings of self-preservation. Children, I don't want to lecture, but you must understand. From the beginning, I offered you my hand to help you experience life. The hand I gave to you is your life and your feelings of self-preservation. And what did you do? You took my hand and put it in a people-made hand puppet that looks like you. That puppet is your red-cubic culture, your so-called civilization. Instead of relating directly to me, you have been relating to me by relating to that hand puppet. That would be fine, but your puppet is dreadfully prejudiced against me, and you have deceived yourselves into believing that the puppet and my hand are the same thing. Even now, you are probably entertaining that notion in the back of your minds, because, if these words appear in a book, they are merely inanimate printer's ink, paper, and symbols. But they are not really me; I am whole life. The puppet expresses only his opinion of me. The real me lies behind the puppet's words and deeds! I am the hand within.

"A hand is an intricately integrated relationship of feelings, fingers, bones, skin, nerves, blood vessels, hormones, minerals, and gases. A hand is animate and alive.

"A puppet is a cultural object that serves to mask the hand that guides it. It is nothing but a coverup of wood, plastic, plaster, glue, and metal. The puppet of your civilization is a controlling inanimate thing, just like a soldier's uniform which often acts differently than the feelings of the person who is wearing it.

"Because the puppet you've placed on my hand looks like your culture and speaks your language, you relate to its stories and believe that you are relating to me. You believe it can make a euphoric womb for you to retreat to—but it can't. A puppet does not have the knowledge of life to make a lasting, supportive womb. A puppet is relatively lifeless—that is what it is teaching you to be, and that is why you are going in that direction.

"But the life force that is hidden underneath the puppet is still my helping hand. I still speak to you in my language of feeling. If you try, you can hear me by listening to your self-preservation feelings. They are me without the puppet. "I wish you wouldn't listen to that blasted puppet. I can hardly ever talk to you directly any more; you pay attention only to the puppet. I'm trying to help you to survive and to obtain wholesome euphoria, and you just hear the puppet, who tells you about the euphoria of your technologies. You'll notice that it never tells you how to survive. When it comes to that, your civilization is not only a puppet, it is also ignorant. The puppet may talk to you, but it knows very little about life.

"Out of its own ignorance, your puppet teaches you that I am out to kill you, and you believe it. That's why the further away from me you get, the better off you think you are. You don't seem to see where it's leading you. I do, and it worries me terribly. Where can you go? I'm your universal mother; living with me is the only home you have.

"Things get even more difficult, because you think you hear me saying, 'Come home and live with me,' and all you can visualize is going home to your human parents and giving up your freedom and your adulthood. You mustn't confuse your two homes or your two mothers. Don't treat us the same. Your home with me is the planet Earth. If you really want to be mature adults, be mature enough in your thinking to accept that, in your relationship with Nature, an adult is a child. You are all my children. Don't confuse your symbols. I am your lifelong mother. Your parents are the mother and father of your immediate childhood.

"From the look on Mike's face, I take it this kind of talk reminds you of the scoldings of your childhoods, and you don't enjoy it. I've cracked your red-cubic walls. Well, I'm sorry, but as adults perhaps you can face the facts of life; you can't live without my allowance, and I can't give it to you, unless I am kept whole so that I can produce it.

"Originally, you were born to Mother Earth (who is a part of me) in one specific part of the globe, maybe East Africa. You seemed to be very well adapted to life there. The problems between us began when you decided to see the world. You traveled to other environments, where you were cold and hungry because you were not suited for those environments.

"Your puppet told you those areas were hostile. It was wrong. Those environments were part of me. They were part of the Earth that made your life possible. Those environments, when you moved into them, shared their life-giving gifts with you. Because I fluctuated differently in those new environments, you had to put more energy into surviving, but you survived everywhere.

"But that didn't satisfy you. You had cut yourselves off from some of my euphoria, and your puppet told you that you had the ability to create your own euphoric environments with your technologies. It told you that in your new environments you could safely ignore me, and create your own womb. You believed the puppet and you believe it still.

"I've been trying to reach you American children for almost four hundred years. I've spoken to you through your feelings, your instincts for life. Even the puppet has indirectly told you of my importance to you. You have heard me in the words of hunter-gatherers, small organic farmers, Henry David Thoreau, Rolling Thunder, Aldo Leopold, John Muir, William O. Douglas, Edward Abbey, Tom Brown, Jr., and many, many others.

"But the puppet tells you on a feeling level that these people who sow my messages are nuts—harmless, but cracked—and you believe that, too. You listen to practically everything your puppet tells you. It breaks my heart.

"If you are not going to listen to my words, listen to your own senses. Look at the problems of your lifestyle. Think about the prospects for your survival, if you continue the way you are going. Then compare your prognosis with a similar evaluation of the lives of those who respect and enjoy me.

"Look at the difference between those people with simpler, less technologically dependent lives, and those people who hide from my fluctuations in physical and mental fortresses. Do this, and you will discover that those who live in harmony with me are not what the puppet tells you; they are what they are: happier, more self-sufficient, more independent, and leading fuller lives. Who is leading you to destruction? Who is killing you, your mother or your puppet?

"I don't want to sound impatient, but of course it's the puppet. That should be easy to accept, but the puppet of your civilization has prejudiced you so thoroughly against me that you can hardly see the plain truth: *it is a puppet and I am life.*

"I've got to talk to you about this prejudice against me. Mike, don't squirm. When we get right down to it, it's a matter of life and death, isn't it?

"You Americans have a lot of feelings about death which

prejudice you against me. I don't know where you got them—not from me—I never even heard of them, until I heard you talking about them. That blasted puppet, most likely.

"Death is part of the adventure of being alive. Death is the more restful inanimate part of the life process; death is the circumstance that allows for new life and for growth. It's part of the life experience. Everybody and everything tries to avoid it when it comes around. I've taught you to take precautions; that's only natural. Death *is* important; but, of course, so is life. Most of your precautions amount to putting up a sign saying, 'Will be back next week,' so when Death comes to the door, he figures nobody's home, and goes elsewhere. That's fine. You both know he'll catch up with you someday, but you don't worry about it. At least that's how it should be.

"But you've picked up some crazy ideas about death. You've got yourselves so turned around that half the time you have yourselves almost convinced that you're not going to die at all. You don't want to accept that in you, as in every other animate organism, there is an end to your growth. The other half the time you're so scared of the death that's waiting for you that you can hardly live, for hiding from it.

"I know a New England cemetery that houses a huge fortress-like mausoleum, covered with sculpture. A rich man is entombed in it, laid out in an immaculate granite coffin inlaid with gold and silver. It's a sight. The protective marble vault is carpeted, electrically heated, air-conditioned, and wired for stereo background music. There is a porcelain bathroom and maid service. Oh, the man's dead all right, but he made provisions in his will for this set-up, this denial of any connectin with me. I was hurt, but even more so when I overheard one of his business acquaintenances, leaving after paying his respects, exclaim enthusiastically, "Man, that's what I call Living."

"How did your ideas about life and death get so scrambled? I can't understand it, and it grieves me so.

"You've accelerated the pace of your living, and in doing so have diluted life's quality. Life is time. When you lived with me, there were no schedules or time clocks. Because there was no consideration of time, you had all the time in the world; you lived life, you *participated*. And when death came, generally, it was quick—and at the very end, it was almost always painless.

"That's one of my kindnesses. You've probably heard of soldiers, horribly wounded—their legs blown away or half an arm shot off—who have no pain, who say they are fine, even as they die. When death calls, I am your anesthesia—but you don't seem to recognize my gift. You even seem to think I'm out to get you. Me, your mother! Kids, I don't enjoy the sight of fear or suffering any more than you do. Too much of either one, and there I am to comfort you, to calm you with my natural sedative, which you call shock.

"Mike, you must remember that foggy day when you fell, climbing in the canyon. You thought it was a simple hike up a gorge that had been cut through a marble deposit. But I had needed rain, and water cascaded down the center of the gorge, painting a striking scene of glistening, polished marble walls, and wet, swirled marble basins. The simple route up the canyon floor was awash with rapids, but you discovered a shelf of rock snaking up the slick side of the gorge.

"With little thought you made your way up the thin ledge. A steep incline of wet rock and slippery footing proved your undoing on the high path above the canyon floor. It all happened very quickly. Your feet shot out from under you, you had a moment of terrible, sliding fear; and then I came to your comfort, you blacked out, with shock. Shock is the loving protection I give to all my children in trouble, that they might feel no harm on their gentle journey home with me.

"And you did not fall, as you had feared. Your unconscious body remained on the ledge, and you recovered seconds later, when I signalled from inside you to announce that you were safe. You did not, that day, shift from one form of life to another. You did not experience the misunderstood life process you call death.

"Perhaps all of you fear living like wildlife, because you fear dying like wildlife. And who would know better than you how vulnerable and exploitable wildlife is? Some of you carry high-technology automatic weapons and murder wildlife for your amusement; to those among you, killing is a sport, a hobby. Not a very reverent attitude towards life, now is it?

"Quick, ultimately painless death is my way in the wild. Of course there is pain in life, but when suffering is too great, I ease it with my tranquilizing shock. The chronic sufferer is soon taken by a predator—and in the end, that too is painless. I do not consign my children to the years of suffering and indignity which your puppet commonly prescribes for your loved ones and your elderly in hospitals and nursing homes.

"You might use your heads a little; that's why I gave them to you. Your indoor lives are limited and unchallenging. They take their toll. It breaks my heart to see you lost in feelings of isolation, depression, anxiety, and helplessness. You've boxed yourselves in and me out, and you can't understand why you get lonely, why in quiet moments you feel empty inside. It makes me sad. Your lives never were meant to be this way.

"You'll never experience the wholeness of life while many of the naturally euphoric qualities with which I infuse it are robbed by your machines.

"Your automobile ride may be quick and weatherproof, but it is restricted to the beaten path—the cement path!—and your driving skills are so automatic you almost can travel in a stupor. There is all the difference in the world between riding two miles down the highway and walking two miles through what remains of your wild inheritance from me. The difference is an awareness of life—your own and those of the life forms around you. The difference is my euphoria.

"I've watched you abuse me; I've come to the conclusion that no one, in good conscience, could do to me the things you've done, unless he or she were strongly, prejudicially misled.

"The puppet of your civilization has misled you. It has taught you to believe in lies—we'll call them myths—about your Mother Nature. I've watched your methodical destruction of our planet, and I've listened to your prejudiced institutions. The prejudiced myths that you were taught by your puppet, and hold most dear, have become obvious to me. Not all of you believe in all of them, but most of you believe in more than one.

The Myth of the Dead Planet. You believe that I and the planet Earth are dead, boring, or harmful.

The Myth of Cultural Euphoria. You believe that you can find the euphoric security of the womb in the fortress of your technologies, the faith of your religions, or the warm daze of your drugs.

The Myth of Cultural Righteousness. You believe people are not Nature; your conditioned, cultural feelings are held to be right, because they are accepted by your civilization as right; your natural self-preservation feelings that I gave you have been conditioned to feel threatening and taboo. You reject the fact that you have two mothers.

The Myth of Status and Power. You believe, no matter if a social setting is destructive to me, that security, status, popularity, and power roles in it are admirable goals for you to pursue.

The Myth of Competition. You believe in competition as a natural phenomenon—which leads you to believe that it is impossible for you to live in equilibrium with me or your neighbors, because you believe that you must compete for resources and survival.

"I suppose I ought to keep those myths in mind when you hurt me—keep them in mind and somehow disregard the hurt—it would be the logical thing to do. But I don't deal in logic; I feelingfully deal in life. My feelings are still hurt, bewildered, and sad.

"Your prejudice against me puts pressures on me that I never knew could exist, and you give me no time at all to begin to figure out how to cope with them. Why can't you lay off your toxic wastes, pollution, habitat destruction, mega-technology, and atomic power for about thirty years. That way, together, we could recycle something neat. We'd be a family again.

"The way things are going right now, each time you hit me— and yourselves—with that junk, it makes both of us less able to deal with the next blow. That's why neither of us are as healthy as we once were. Parts of us are still reeling and recovering. And your puppet, your so-called civilization, keeps prejudicially ticking along, making things worse. As I try to help you live, it tries to make you into puppets controlled by itself.

"For example, take the media.

"You live in the media. You read the paper over your breakfast coffee; you drive to work listening to the radio; you watch television during breaks; you flip through the magazines in your mail when you get home; and after dinner, you go out to a movie.

"In the latest eighty-page issue of *TIME* magazine, the reader is immersed in murder, war, terror, technology, science, pollution, bombs, and alcohol... Maybe you'll find it more understandable in terms of your math and numbers.

"The cover article is about the drug cocaine, a thirty billion dollar-a-year underground industry. Fifty per cent of the same issue is devoted to advertising, of which one-third hawks the national poisons: liquor and cigarettes. Your mass media *is* pollution. One-third of your waking hours are spent with television and radio and newspapers! As if that weren't enough, your buildings and roadsides—yes, even your clothing—all are covered with billboards selling false euphoria. All those advertisements are designed to appeal to the myths imbedded in your inner consciousness-usually the **Myth of Cultural Euphoria** and the **Myth of Status and Power.** They trigger off your anti-Nature

conditioning and exploit it. And you, in turn, exploit me.
"That media stuff is violently prejudiced against me, but you
can't see it. Most of you have been surrounded by it, since the day
you were born—you don't think about it twice. But it influences
you.

"Take your space shuttle. Columbia, you call it? That thing is a
great accomplishment, technically, but when you rocketed it up,
you surely put me down. That rocket shuttle is a huge flight of
technology and an ultimate flight from me. You're not escaping me
or your problems in that shuttle, you're taking them with you,
you're carrying them in your heads and hearts. Competition, envy,
greed, exploitation, hate, destruction. Why, you predict them in
your own science fiction film fantasies. You go to see *Star Wars*.
Wars! You hate me and love wars—you can't wait to line up and pay
your money to 'enjoy' *Star Wars*.

"Your space shuttle is a technological summation of all your
prejudicial myths and thinking. Just hear what your media had to
say about it; look how they reinforce your prejudicial myths
against me:

> We can live in outer space and create our own eco-
> system away from Earth. (You can create your own
> technological womb.)
>
> We are highly successful. (Status, popularity, and
> power roles are admirable; you must compete and
> be winners.)
>
> We are exuberant about this giant piece of
> technology. (Womb euphoria can be found in
> technology.)
>
> We have reached the height of excellence. (Competi-
> tion and status.)
>
> We have outcompeted the Russians. (Competition
> and status.)
>
> We have increased our military power. (Competi-
> tion instead of cooperation.)
>
> We have conquered Nature in space. (I'm harmful
> and must be subdued.)
>
> This emotionally makes up for our defeat in
> Vietnam, Iran, and Three Mile Island. (Competition
> and winning rather than cooperation.)
>
> Its flaming, powerful rockets lifted us out of our

collective sense of gloom. (Technology provides euphoria.)

Astronauts, through you we feel as giants again. (Power over me triggers euphoria in you.)

We can now control the high ground of space. (Competition, power over me.)

It shows everybody that the U.S. is still number one. (Competition and status.)

A retreat from Earth's hard realities. (Womb euphoria in technology.)

It's a touch of euphoria. (Right on—but way off.)

The militarization of space is hailed. (Competition, power.)

A reaffirmation of American technological prowess. (Technological euphoria.)

Shows the world that "Americans still have the know-how, and Americans still have the true grit that conquered a savage wilderness."

—President RONALD REAGAN

"It makes me sick at heart. The ten billion-dollar cost of the space shuttle, if applied to improving your quality of life on Earth, would eliminate your expensive urge to leave the mess you've made of our planet...but you never consider it.

"You'll never fill your childish womb needs this way. Only three months later your media was telling you:

> The United States needs a new shot of pride to rev up her lagging technology. A failure of the U.S. to technologically compete effectively could have grave consequences for the national psyche.

"What a treadmill you've created for yourselves. And you go on to build new and bigger treadmills. Maybe you should come home for a rest; perhaps when you're feeling better you can clean up your room. If you stay where you are, your treadmills can only make you sicker. Think about it. That's what the tobacco industry is doing to you.

"The tobacco industry is feeding you death-dealing media messages and you're eating them. You've been away from me so long, many of you can't hear my voice in your self-preservation feelings. Those tobacco people are manipulating you so cleverly, you don't seem to remember that life is worth living and that they're stealing it from you by selling you poisons. They are killing everything I taught you about self-preservation. Such crap:

> 'Alive with pleasure.' 'The low tar that won't leave you hungry for taste.' 'The pleasure is back.' '10 packs of Carlton have less tar than one pack of Camels. Carlton is lowest.' 'Marlboro Lights. The spirit of Marlboro—' (the macho, western cowboy outdoorsman). '—in a low tar cigarette.' 'You've come a long way, baby.' 'You found it.'

"That cigarette advertising is appealing to your desire for the status, identity, pleasure, spirit, and tastes that you lost when you left me out of your lives. The glossy photographs of wild areas and beautiful people in the ads hint at a connection between relationships, desirability, sex, excitement, your Mother Nature, and— *cigarettes*. They suggest to you that you can't enjoy the former without smoking the latter. And if that fails, the word 'sex' is often written subliminally on the skin and hair of the models in order to hold your attention. It's criminal.

"It pains me to see you hurting yourselves this way. Cigarettes are obnoxious poisons that are the principal cause of preventable chronic disease in this country. Even your numbers tell you that much: seventy-five to eighty per cent of all lung cancer, emphysema, and chronic bronchitis, as well as thirty per cent of coronary heart disease is caused by cigarettes. You've heard these figures; you know the dangers—you even read them on every pack. But you don't learn. The objectivity of your numbers and percentages and scientific studies seems to allow you to disregard their message. Well. I'd take those cigarettes and make you smoke them the way they affect you. I'd have you smoke them with the lit ends in your mouths. You'd get burned, but you'd learn. ...A mother's love will try anything to keep her children safe and sound. Can't you see you're nothing but a piece of meat with money as far as those tobacco guys are concerned?

"Your prejudiced media attempts to lure you away from me in other ways. They seduce you with the luxury of womb-euphoric products and lifestyles like this:

> Find your quiet corner at a castle. Aye, a castle, a spectacular, self-contained resort overlooking a mile of Caribbean beach. A stunning combination of 19th century elegance on a tropical seventy-two acre estate, and 20th century amusements. Seven lighted tennis courts, a pro shop, three pools, free

movies, shuffleboard, a starlit disco, barbecues, billiards, live nightly entertainment, rum parties, moonlight cruises, three restaurants, a beach cafe, terraced accommodations, air-conditioning, gaming tables, plush casino, native cuisine, sailing, golfing, rafting, perfection, romantic rain forests, bright bikini sunshine, people with laughter in their eyes, and a cool prevailing sea breeze.

The thunder's still there, that unmistakeable reflection of the distinctive and dating heritage. The lines are crisp, the stance bold and assured. Inside, the plush interiors and electronic magic create a distinctive driving environment. Thunderbird drivers appreciate style and flair and expect it. It's good to know the thunder's still there.

I'm lonely, I'm yearning for you to come live in me and share my attributes. I'm classic, bold, and cozy, a Victorian home restored and modernized to royal taste par excellence. I have three bedrooms, three full baths, a large kitchen, living and dining room, a vestibule with a built-in bar, a sewing room, wall to wall carpeting throughout. I'm tastefully designed with an outdoor deck; built-in 6-foot TV-stereo-radio-tape combination; have a workshop; sunken bath with Jacuzzi; laundry room fully equipped with washer, dryer, and steam press; electric heat and hot water; air-conditioning; dehumidifier and dust eradicator. I also provide you with a gaming room (billiards and electronic games) a classic library and study; automatic insect and rodent control; and automatic electronic built-in burglar, fire, and freezing alarms that are connected to the police and fire departments. I am surrounded by four acres of impeccably lawned grounds with both swimming pool and tennis court. Please come and visit; to know me is to love me. Satisfy my needs and yours.

"You can see that the media are simply your written and spoken symbols, and most of those symbols are prejudiced against my fluctuating pulse. Believing and acting on symbols that are prejudiced against me is as stupid as being prejudiced against your own heartbeat.

"Yet historically, some of your most distinguished forefathers' minds have been extremely prejudiced against me. Perhaps they were heard only because you were so eager to listen to them—for you ignored many others through whom I was speaking.

"Thomas Hobbes thought that living harmoniously with me as a hunter-gatherer was bound to be *a solitary, poor, nasty, brutish, and short affair.*

"Francis Bacon told you in the mid-17th century that the pursuit of objective knowledge would allow you *to take command over things natural—over bodies, medicine, mechanical powers, and infinite others of this kind.* That same year the Pilgrims landed at Plymouth. Bacon was saying that you should be objective and ignore your natural feelings; in other words, ignore me. You did—and Cape Cod has never regained her forests.

"Then Descartes convinced you that mathematics is the most powerful instrument of knowledge and is the source of all things. That screwball had you believing that only space, matter and motion are important, that those are all that I consist of, and that his blasted mathematics represents the total order of things. Descartes eliminated my heartbeat; he conceived a euphoric universe gained through the sterile mechanics of mathematics. I disliked that bigot.

"Then Newton came into the picture. Dear old Isaac. He reduced me to three laws of motion. Fluctuating, disorderly, colorful, smelly, oozy, tasty me! I became in your minds an orderly, mechanical machine that ran on mathematic formulae and scientific observation.

"And then along comes the Adolph Hitler of Nature, John Locke. Locke declared the *negation of nature!* was the way toward happiness. I could scream! He told you that self-interest was the only basis for society, and that the purpose of society was to increase the property of its members. Locke called the land that was left to me *waste land!* He wasn't much more popular with me than Descartes.

"Finally old Adam Smith entered the you-can-find-euphoria-in-materialism scene. Smith drew up laws based on materialistic self-interest; he said that the pursuit of economic gain was a virtuous activity and a natural force. I cherished him like an Arab is cherished in Miami Beach.

"All these characters spoke for the puppet of your civilization, and they fill your textbooks and language. They were the womb addicts that built the cradle of American culture by tearing me apart.

"It's still going on today, you know. People still insist on dividing me up into laws and matter. The latest kick is the second law of thermodynamics. Entropy. They say it's the supreme law of Nature; basically it's nothing but one of your physical laws that says I'm running down, as any living system will. Now that may be true, but how do you explain that newer life forms are more complex and that the sun's energy may still be increasing? I could actually be raising your allowance. Well, I'm alive. I'm living and dying, growing and shedding. I have billions of years to go, and just now what's really running me down and depressing me is your determined use of discriminatory symbols.

"Your feelings are your connection to the pulsating life process of the planet. They are an ultimate truth of life, a language that all other life forms and systems speak. But your acculturation leads you to believe that your symbols, not your feelings are the truth. For example, *True or False:* 2 + 2 = 4? The correct answer is *False* because the basic premise of mathematics is that the number one is a reality. Believe me, that's nonsense! There's no such thing as one.

What is one tree? At the moment you select and describe one tree, you're speaking about that tree *as it exists in me at that moment.* But I've made that tree different from any other tree. And a moment later, that tree has changed because it has grown. It has reacted to my T-R fluctuations of food, climate, weather stress and time. Your culture considers such changes as an insignificant matter. No. They are a matter of survival because they're the essence of life.

It may seem unbelievable to you, but one tree, one anything in nature never really exists. By the time one entity is symbolized, and it is communicated what that one entity was, it no longer is. It's changed. One is false.

What's true is that your symbolization process is more important to you than your feelings. The latter are me, your inner nature. Maybe you love your symbols because, like the number one, they seldom change. They're a "security". Our mutual problem is that I speak fluctuating T-R sensations; you speak symbols. I demonstrate how the world works and you hear me to be a dictator who's denying you your freedom to be detrimentally exploitive. I'm not a fascist. I'm a living organism protecting my life and yours.

Why don't you evaluate your truths by their long-term effects upon me? Then we'd have more in common. Stop hiding from my fluctuations and instead love them. What's a mother to do?"

I HAD A LOT TO THINK ABOUT.

Chapter Eleven

WHAT PROFITS A MAN?

*One wonders if our immorality to each other begins at the interface
with Nature's far-reaching shoreline.*

The glowing, pre-dawn sky silhouetted a cactus that had stood silent
vigil beside me throughout the night. Venus and Scorpio began to fade to
the sun as it glimmered and rose on the desert. The thorns adorning my
cactus cathedral luminesced; gently the land erupted into color where
greyness had prevailed. The sunlight warmed my sleeping bag. Wisps of
vapor, miniature clouds of condensation, rose from the nylon as the sheen
of frost covering the bag melted, ran, and evaporated. The tension of the
chilly night was slowly released and began its journey toward the thermal
climax of midafternoon.

Serena, a student who had volunteered to be the Expedition alarm
clock, appeared and quietly announced, "Time to get up." My wife Diana
and our dog Timber slipped out of our sleeping bag and made their way to
the breakfast area, leaving me alone to pack the bag and watch the day
unfold. (Yes, the dog sleeps with us, or on us, depending upon the wind-
chill factor and the proximity of thunder storms.)

"What a fascinating, glorious bedroom," I mused to myself. "Far more
interesting than that expensive interior decorator job in my New York City
apartment fifteen years ago...cheaper, too."

"I see you're still at it, dear," interrupted Mother Nature, a bit
rudely, I thought.

I responded with the distance in my feelings that her intrusion had
triggered. "You seem aggravated with me," I said.

"Just giving you an honest reaction to your comparison of my
desert dawn with an apartment and its economics," she replied. "I
dislike being associated with the indoor wombs that you have built
and overbuilt to remove yourselves from me. I guess what you are
hearing from me is a mixture of jealousy, loneliness, and resentment. It's shocking to discover that I am being left out even when
you are two thousand miles and fifteen years away from your old
apartment."

I was stung. "You're overreacting," I retorted. "I can't help being what
I am—I can't help what has happened to me. It's the way I was raised, it's
burned into my memory and attached to my stream of consciousness. I am

not responsible for what I was born into, and I'm trying to change."

"You're trying, all right," she said with humor, and then revised her tone, murmuring, "I guess you're right...Thinking about it, I recognize that you *are* making an attempt. It's those who don't see and don't care that I'm really annoyed with. Sorry; I suppose I took my frustration out on you because you look just like them—and a part of you still thinks just like them."

MC: That's prejudice, you know.

MN: Yes. ...I should know.   ·

MC: And I don't think that the rest of the nation doesn't care. I think that deep down they do care, they care very much. It's just covered up by where they are and what they are consistently exposed to, day and night. I think they do love you—but they can't find you.

MN: I try to let them know I still exist. I try to reach them, especially when they're asleep and they've turned off the media stimulants and slowed the pace of that puppet world they've built for themselves. I gave you sleep so that you could lose yourselves in me, recharge your energies, and get to know me better...but sleep isn't serving its proper function for those of you who live so far away from me. Mike, I can't reach them even at night. I have to gouge my way through a whole department store of your daily activities in order to reach you.

MC: I know; it's very hard. You don't come to us alone and pure as you once did. Even when we're sleeping, you share the stage with the puppet. Our dreams, believe me, can be very confusing. Without our daytime sensibilities to arrange things as we were taught, everything seems crazy. You come to us as you really are: one wild and free feeling after another. But the puppet is there, and those feelings are attached to puppet actions that don't make a bit of sense. What a wild ride—sometimes it's very disturbing, even scary—I've woken up screaming and punching my pillow.

MN: Sorry about that. I can only do what I can to break through the mask of the puppet you have placed on me. I guess that is what happens when you take life for granted.

MC: Wait a minute, Mother. We live, we don't take life for granted—we experience it all the time.

MN: What you call life and what I call life are almost entirely different experiences. Why, just take sleeping. We're only speaking the same language when you have a sound, restful sleep that lets you awaken relaxed, refreshed, and looking forward to relating to my changeable whims. That's a rare occurence for most of you—yet for many of my other children, it is a daily and nightly experience.

MC: How 'bout when the dog yelps in his dreams at night?

MN: Depends upon how acculturated you've made the dog. Ofttimes he's dreaming of life with me, chasing a rabbit or rolling happily in the protective smell of a dead animal. But you're correct, sometimes he's reliving a scolding he's received—but that's a different cry then, isn't it?...I could echo it; most of you treat me like a dog and worse.

MC: We can't help it; we've been taught to live in houses, cars, offices, schools. We've grown up with the media.

MN: But you are grown up, now, and free to choose.

MC: What choice do we have?

MN: (irritably) You could start with your own back yards. It hurts me terribly to see those of you with lawns spend huge amounts of money, chemicals, time, and energy to make and maintain them. Lawns could be wild areas that leave room for me. Instead, and enforced by zoning laws that inflate land values, your lawns pollute water systems with chemicals and draw energy with mowers. You could have birds, wildlife, meadows, trees, and wildflowers instead of grass rugs to mow and crabgrass to weed. What's so terrible about having a habitat instead of a headache?

MC: Nothing. But you said, yourself—lawns are a law in most places.

MN: Yes. That kind of law is prejudiced against me; it's a fulfillment of many of your myths about me. Are there so few among you that have the integrity and the guts to confront one little law, to show that it is an uncivil law as far as your relationship with me is concerned?

MC: Well...

MN: Frankly, that's why you have bad dreams. Your day-to-day interactions with me are a nightmare, but your puppet won't let you see it. At night, though, when the puppet's authority is relaxed by your sleeping, I can reach you through your self-preservation feelings. Sometimes your dreams upset you and scare you, because your self-preservation feelings manage to break through the constraints of the sleeping puppet and tell you what your nature knows to be true.

MC: Which is?

MN: They come very close to letting you know just how you are hurting yourselves, killing yourselves, in your refusal to be strong and forceful about your self-preservation and your desire for a healthy relationship with me.

MC: Mother, you say our daily lives are nightmares. That sounds pretty strange to me; I bet it would to a lot of people. Most people enjoy their daily lives, and certainly wouldn't liken them to a nightmare.

MN: That's exactly the point. You not only accept and defend the destructive materialistic euphoria of your daily lives, you accept and defend your right to have the nightmares it brings in its wake. You don't seem too concerned by how badly you may be feeling day or night; you're more concerned with what and how much you have. Your nightmares are a reflection of your rotten relationship with me.

MC: Why do you want to give us nightmares?

MN: I don't want to give them to you, I'm just trying to wake you up to the reality of the consequences of your treatment of me. It's too bad I can reach you only at night, but your equivalent daytime nightmares—depression alcoholism, suicide, crime, war, pollution, deliquency—don't seem to make an impression on you.

MC: This is a lot to take in at once. I have to think.

MN: (smiling) No, you have to feel. But go ahead. And I must say, Mike, I appreciate it that you want to hear my sentiments. So few do, these days.

MC: (slowly) Okay. I'm think ——no, *feeling* the truth of what you're telling me. But can you give me examples of more subtle negative treatment of you? It's still hazy in my mind.

MN: I felt hurt when a man bought a store and developed it to be six times larger than it had been. To accommodate the antici-pated customers, he paved over my birches and wildflowers for a parking lot and delivery area. An advertising campaign and discount sales brought in new customers, but there had been no population increase to warrant the enlargement—it was done for profit only. Mike, in America alone, a square of land seventy miles—*seventy miles!*—on each side is being lost every year because you allow yourselves to be manipulated into sterile, impersonal, gargantuan shopping malls, high-ways, developments, and the like. Try to understand: your sticker prices and price tags do not reflect the true cost of your standard of living; you are not paying for it with your cash. You are paying for it with me. And when you need me or those paved-over habitats for supporting life—for doing what they do best—they and I just won't be there to help. It upsets me. It's as though my life—and, ultimately, your lives—just don't matter.

MC: Well, I wouldn't say that...

MN: (sweeping on) And your stores, malls, and industrial develop-
      ments are no different from all your power dams. Take a look
      at your northern Maine hydroelectric project.

      Those dams would provide peak power for New England
      urban districts, but you're going to flood one hundred and
      twenty-five square miles to get it. You'll drown the river, the
      brooks and streams that feed it, the banks and shoreline, the
      forests and the undergrowth, the rare species and all the other
      animals and life forms in the area. All of that, a part of me, you
      are willing—and have the technological power—to sacrifice,
      all so you can turn on your lights, your air conditioning, and
      your ovens at the same time. —You're not saying much, Mike.

MC: ...No.

MN: Well, look at the bigotry, the prejudice against me, that you
      Americans manifested when you recently developed thirty
      miles of new roads in a National Park so that people could
      drive instead of walk to the exciting natural areas. 'If we didn't
      do this,' explained the Park Superintendent to the press, 'many
      people would not feel the park had any value and would not
      support it. We've sacrificed two hundred acres of wild land to
      satisfy them.' He was probably accurate—and it hurts me.

MC: I would think that you'd like having people learn about you in a
      park.

MN: Mike, you finally get together enough sense to preserve a
      place for me so that you can begin to get to know me again, and
      what do you do? You tear it apart so you won't have to walk.
      Walking is natural, walking is me. I just don't understand.
      While you're teaching about my values you are teaching that it
      is okay to tear me apart. Why, you're broadcasting the same
      double message that is given in indoor Nature classes, where
      they preach about the web of life and yet capture or kill wild
      specimens for study, dividing my wholeness and destroying
      the symmetry of the web. Both the park and the Nature classes
      in these examples are teaching prejudice—but they don't see
      it.

MC: I can see how that would be very disheartening.

MN: I'm also disheartened to hear that one mile from the park you
      plan to install an oil refinery that will pollute the air and water
      and make the view from the park include supertankers and the
      grime of industry. You people don't seem to realize that the
      coast is an exceptional feeding ground for birds, marine mam-

mals, fish, and other wildlife. I specifically made that area rich
in food, with the flushing action of the tides in local bays. It
supported you and whales and porpoises—but now you want
supertankers to motor in those swift, rocky narrows. You
want to stop the waters of the bays with hydroelectric dams.
You want to build a factory that will hire people who will earn
more money to buy *more* technologies to gain *further* distance
from me, to find the euphoria promised by the media. Will you
ever again appreciate just taking enough to live, and then
returning it?

MC: How could that be done? Is it possible?

MN: There is land which could be harmoniously farmed and
could, at the same time, act as a habitat for what you call
*wild*life; but to farm it harmoniously, you must farm it by
imitating my ways rather than inventing your own. Your own
ways are impatient. They don't wait to recycle. They seek to
regain the immediate euphoria of the human womb; my ways
impart the euphoria of the living womb of the planet. You
always forget that. Your human womb is gone once you are
born. If you could accept that, our relationship would be much
better off. Fight your prejudice. You have forgotten that we
have a common interest: life.

MC: Mother, that's foolish. Of course people want to live.

MN: Aren't you killing me? How can you even begin to think you
want to live, when I alone am life and you are destroying me?
The fish populations in your lakes are dying off because of the
acidity from the rainwater. You make acid rain and unhealthy
breathing with your industrial smoke—that's prejudice
against life. It's a technological threat to me that is ignored by
your industry unless it is confronted by the government—and
the government is largely controlled by the business community. Where is life in that cycle? Who amongst you represents
me and my interests? Who speaks for life?

MC: Our government was founded, 'of the people, by the people, *for* the
people.' You aren't a person.

MN: Oh, shame on your narrow, bigoted mind! What a terrible
thing to say. That's the epitome of discrimination; it's really
painful. Don't you realize that you give the legal power and
identity of a living person to your collective drive to do business? You allow industry to *incorporate*—meaning that you
endow it with the rights and privileges of a person—even
though it is merely a community phenomenon, a desire to

service or exploit.

MC: (confused) But isn't business a way for people to stay alive?

MN: Of course, in your civilization, it is, but it's often death for me. And in the business of life, you're such terrible business-people that you not only can't maintain your economy, but you're quickly putting yourselves out of the picture.

MC: Why is the economy—and our prospects for life in the future—so unstable and uncertain?

MN: You create a vicious circle of prejudice against me, that's why. For example, haven't you noticed that government services and spending, designed to increase distance from me, have increased the need for raising property taxes? This hurts me and it's bad for you; it forces property owners to subdivide and sell off parcels to meet their taxes. The subdivided parcels often go for housing, which in turn, increases the demand for government services and spending. That's just one of many examples of what happens when you give in to prejudice against me. —You try to gain distance from your natural womb. The more effort you put into doing it, the more distance you gain, the longer and more complicated is your supply line, the greater are your expenses and loss, the more bewildered you feel, and the harder it is for me to help you stay alive. That's not good business; that's suicide.

MC: Suicide!

MN: Your government fights me and I'm not represented in it. The taxpayers demand a budget cut and the first cut removes environmental education from the school curriculum, because *Nature study is a frill that can't be afforded.* What am I to do? I'm trying to help you, and instead of listening and learning how I can help, you cast me out. Learning about me is per-ceived as a threatening waste of time in your materialistic, consumer-oriented society. Of course, the underlying threat from me is that some of you might attempt to become indepen-dent and self-sufficient, if you discovered what is being done to your life systems, and thus to your own lives.

MC: That's not completely true...

MN: True? You have no idea what truth is. The truth is that you are always seeking the truth so that you won't have to ques-tion truths, once they are found—once you have agreed upon them. Truth is a resting place from life, from tensions, ques-tioning, adapting, and relating. Truth is static, it's the name your culture gives to womb-like understandings and relation-

ships. After all these years you'd think that you'd have caught
on that your search for truth is actually a search for a permanent escape from my fluctuating pulse. You want to hear the
truth? You are talking to Truth right now.

MC: You are Truth?

MN: You are prejudiced against the truth of your own Nature,
and the truth of your lifetime relationship with me as an
embryo in the womb of the Earth. Your cultural truths bring
only short-term euphorias. None of those truths have held up
historically except those which accept me for being me. You
Americans dislike me so much that you refuse to accept the
truth that I am your mother—yesterday, today, and tomorrow.

MC: You make us sound so inhuman, so spoiled, cold, and calculating.
That's unfair.

MN: Oh, stop that whining. You've always appeared that way to
anyone against whom you were prejudiced. Ask Blacks,
Indians, minorities, foreigners, women. You keep trying to
avoid the fact that you are prejudiced against me. Now you're
saying that you're human, and humane. Humane, indeed. You
may at times be humane to your families and your customers,
but when it comes to me, your natural mother—why, you hunt
baby harp seals and skin them alive before they lose
consciousness. Violence has become part of you, you find it
exciting. It is used to liven up otherwise dull or plotless
movies and books. You have used and do use violence against
me; read this:

> Two hundred miles off the coast the gun roars. There is
> a momentary silence and then a muffled explosion as the
> time fuse functions and the grenade fragments deep inside
> the whale. Then follows a fight to the death between the
> mammal and the crew of the catching vessel. The whale is
> doomed. It is hit by a second harpoon, causing a dull
> explosion in its vitals. Then comes a series of
> convulsions—a last desperate struggle. The whale spurts
> blood, keels over slowly, and floats belly upward. The last
> remaining members of my irreplaceable species are being
> killed for candle wax, for soap, oil, pet food, margarine,
> fertilizer, and perfume. Two million whales have been
> killed in the past fifty years.*

---

*Adapted from *Call to the South,* by R.D. MCLAUGHLIN, London:
Harrop & Co.

Was that humane? Did you learn anything from it? Has it made you more human, or more powerful?

MC: That's not an accurate picture of us. We've stopped doing that. Anyway, read the Ten Commandments, look at our laws, all our law enforcement agencies; they're all against violence.

MN: They are against people's violence and inhumanity to people, not to Nature. Your laws are your attempt to stop the inhumanity that has become habitual with you as a result of your prejudicial relationship with me. Do you think the hunter-gatherers needed such laws?

MC: I suppose not, but—

MN: You people refuse to recognize all life as equal, with an equal capacity for suffering—a refusal which allows your conscience license to inflict pain and cruelty. The unemotional, objective viewpoint of your *science* inadvertently condones insensitivity and brutality. It is hard to understand why you hold scientists in such high esteem. Many of them practice being inhuman. They don't even speak the same language as you do; their elitist terminology removes them from the words that would symbolize for them pain, torment, and their inhumane relationship with Nature. To some scientists, suffering has become merely *behavioral patterns* and *positive* or *negative reactions*.

MC: I'm sure you're exaggerating, Mother.

MN: Yes? How then can you explain these tortures, that are being practiced right now with minor variations, across the country? Read, Mike: it's part of something I was encouraging Peter Singer to write about.

> At your universities, dogs receive psychological experimental training in shuttleboxes where they are electrically shocked. The dogs run frantically about, defecating, urinating, and howling. Hundreds of intense electric shocks are delivered to the dogs' feet through a grid floor. Initially, dogs show symptoms such as yelping, shrieking, trembling, and attacking the apparatus. The university experiments are conducted to observe dogs' reactions to shock and their avoidance behavior.
>
> In your laboratories, rabbits held in neck stocks have eyes ulcerated. The eyes are badly inflamed, open, running sores which will deteriorate until only half the eye is visible. In a few days the eyes will become blind as they are literally burned out of

their sockets. The rabbits are being used in tests
that consist of dripping into the eyes concentrated
solutions of different substances. Swelling, red-
ness, destruction of iris or cornea, loss of vision, all
are measured, and the solutions' eye irritancy scientifically established. Rabbits are used because they
do not have tear ducts and so cannot flush irritants
from their eyes or even dilute them. The irritants are
new varieties of cosmetics: eye makeup, face cream,
hand lotion, toothpaste, mouthwash, shampoo, hair
conditioner, perfume. (All the stuff you slather on
yourselves in order not to look as I made you look.)
Their eye irritancy capacity in humans is being
measured against the effects on unanesthetized
rabbits.

Psychological investigators have a fascinating
idea: *to induce depression by allowing baby monkeys to attach to cloth surrogate mothers who can
become monsters.* The first of these monsters is a
cloth monkey mother, which ejects high-pressure,
compressed air. It blows the baby's skin almost off
its body. Another *mother* rocks its baby so violently
that the baby's head and teeth rattle, while a third
mother ejects the baby from its ventral surface. The
fourth *mother* ejects sharp brass spikes into its
infant. In all cases, the infant returns or holds
tighter to it's *mother.* In a corollary experiment, three
real female monkeys are made psychotic by raising
them in complete isolation. The monkeys are then
impregnated, and when their babies are born they
smash the infants' faces to the floor and rub them
back and forth. *To induce depression...* If you
respected me, you wouldn't have to worry about
being depressed.*

Your pet stores sell monkeys. Your laboratories
and zoos also buy them. Far away, a hunter shoots;
the female gibbon, mortally wounded, clings to life,
while her baby still clings to the long hair of her left
thigh. The mother misses the branch to which she
leaps, and in a final, desperate effort, manages to
grasp a lower one. Her strength is ebbing away, and
she is unable to pull herself up. Slowly she weakens; her fingers begin to lose their grip. Death is

*Adapted from *Animal Liberation,* by PETER SINGER, New York: Avon
Books, 1976

there, staining her pale fur. The youngster flattens
himself in terror against her bloody flank. Then—
the giddy plunge of over one hundred feet, broken
by a terrible rebound off a tree trunk. The hunter's
object is to capture the baby for a zoo, but his exercise has been in vain: the baby's neck is broken by
the fall.*

A listing of experiments you people do on my animals in the
name of science, research, and medicine would include: accel-
eration, induced aggression, asphyxiation, blinding, burning,
centrifuging, compression, concussion, crowding, drowning,
drug testing, experimentally induced neuroses, freezing, heating,
hemorrhaging, hind limb beating, isolation, multiple injuries,
protein deprivation, punishment, radiation, starvation, shock,
stress, thirst, and more. Unanesthetized animals have suf-
fered and died from these causes, and you scientists have
watched unemotionally and made notes. It's disgusting.

MC: I can understand why you're upset with some of our scientists, but
most of us don't even know that those things are happening.

MN: That's because your culture accepts that they happen; those
experiments are just an aspect of your puppet, just another
facet. Those things are not news, so the news ignores them. If
they became news, why, they might make you upset. Your
feelings about self-preservation and your love for me might
actually come to the surface. That's why you finally protected
whales; they were no longer profitable. Your news
organizations were forced to pay attention to cruelty which
had been going on for centuries and to make you aware of it.
But generally your society does not want awareness, because
the constraints of your emotions would limit the economy;
stifle free enterprise; check your freedom to abuse me as much
as you can.

MC: Why do you think we treat you this way?

MN: Because you take your good feelings for granted. How do
you think it feels to be a stone on the surface of the moon—an
inanimate object on an inanimate celestial body? Do you think
that good feelings—as you know them—exist there?

MC: Well, uhh...probably not. I guess it would be like what we call
death.

*Adapted from *The Animal Connection: The confessions of a Wild
Animal Trafficker,* by Jean Yves Domalain, New York: William Morrow
& Co., 1977.

MN: That's what I mean. You people forget that it is being animately alive that gives you your good feelings, and that I provide your animate life. You often take your good feelings and believe that you have made them through technology, while you take your tense feelings and blame them on me. That's not right. You experience your good feelings because you are alive—and life is me.

I am the part of you that is your feeling about the underdog and the oppressed, like *Bambi*, E.T., and *Watership Down*. Because your puppet is prejudiced against me, when it comes to feeling the value of yourself, *I am outside of your red-cubic consciousness.* You haven't learned to let yourself experience the value of your life-loving feelings about your own life. Instead, you are encouraged to express these feelings towards others who are in trouble. I'm like an unnoticed orphan in your consciousness because you don't fully recognize that I'm a valuable part of you.

MC: Do you have any idea what can be done to change—and correct— our attitudes?

MN: I suppose what I need is a public relations expert to represent me in every aspect of your media, and a lawyer to represent me in every branch of your schools and your legal system, on each of your trespasses against me. They would insist that you relate to me holistically, and that factor might begin to make a difference. Do you realize that right now I am treated legally as if I am guilty first and I must be proven innocent or I will be punished by development or exploitation? I want equal rights!

MC: It sounds like you want to be a corporation with all the rights thereof.

MN: It might be a help—at least I'd get your attention.

MC: Uh—it's an interesting idea. If you were a corporation, what would you do?

MN: Hmmm...You shouldn't have asked that; we might be here a while. Basically, Mike, I'd confront every act of prejudice against me until harmony was reached. There's a long list, you know. Nuclear power, chemical sprays, habitat destruction, materialistic education, overpopulation, techno-luxury, poisoning of predators—and those are the big things.

MC: I'd be interested in hearing more specifics, especially those trespasses which we don't recognize easily.

MN: I'll try. Frankly, it irritates me to think about all this. Okay: I would confront corporation and business pressures that insert themselves into the educational picture and lead students to think that what they are learning at school is good citizenship and survival. I would show students that they are lapping up the business mentality even as they sit with their textbooks in classrooms. I would point out that they are learning the indoor, away from Nature, school atmosphere experientially, so that it really sticks. Competition, cheating, drugs, peer pressure, obeying orders, living in an authoritarian environment. Exposure to these pressures creates people who are perfect targets for exploitation by business. Finally, I would expose to students the myths that underlie their prejudice.

MC: (simply) How?

MN: I would bring an anti-trust suit against the U.S. Department of Education. I'd file over a case I know of where a man with a biology degree was given a job mixing paint in a paint factory, a job that had been refused to a man who actually had been a house painter but who never had finished high school. See, I notice that many jobs are available only to those who have high school or college degrees—though many of the jobs are in fields entirely unrelated to the courses the applicant majored in while in school. That's prejudice against me.

MC: I can't see how that even affects you.

MN: Think about it. Your schooling is in some ways nothing but a red-cubic test. It is a trial to discover which of you are most willing and best able to take orders and tolerate the authority and pressures imposed by the extreme distance from Nature that your competitive, technological positions often demand. Your school system mostly trains, identifies, and congratulates students who have exhibited proficiency in dividing and conquering me. My anti-trust suit would expose this symbiotic relationship between industry and education.

MC: What would that do?

MN: Look. How many times have I seen fishermen operating electronic sonar and other high technology equipment to locate schools of fish? They find the schools—and they proceed to fish them out of existence and themselves out of a livelihood. Why didn't school teach them not to do that? Wouldn't we all be better off if it had? Your educational system has crippled whole communities by making people dependent upon power and authority. Rural people often go on

welfare and food stamps when a factory closes, though their
farm land lies idle and overgrown. This sort of mentality
prevails because self-sufficient farming is not profitable to big
business and thus is not offered as a viable option by the
business-backed media and schools which form public opinion. Your educational system teaches dependence upon institutions and impersonal authority.

MC: Is that what you think I'm doing with expedition education?

MN: Always searching for a compliment, aren't you? No, I don't
think that's what you're teaching. Your program is based on
whole, full-spectrum relationships. I have finally managed to
get it through your thick skull that while you can divide up a
loaf of bread, and benefit from doing so, you can't divide up a
living system without hurting or destroying it. You seem to
have that pretty well understood—we are in tune there.

MC: Thank you.

MN: I also used to be in harmony with Indians and a few settlers
and their music reflected it; their folksongs recognized me, at
least. But your modern songs are a farce. *Hold me so close and
hold me tight, kiss all the hurt of this world away.* That's the
theme to thousands of songs on the airwaves. If your songs
don't extoll the pleasures and problems of drugged or drunken
euphoria, they convey *I feel like half a person without you.*
People are goaded into having relationships only with each
other and finding fulfillment only in romance. The
relationships are unwholesome—as the increasing divorce
rate shows.

MC: (surprised) You mean you don't think that people should relate to
each other romantically?

MN: Not at all. I only mean they can't relate closely to each other,
realistically, if this is done to the exclusion of me, because then
the relationship is only a euphoric escape and a cop-out—a
refuge from the tensions produced by your distance from me.
You people and I, we've been together in America for over four
hundred years. I am the source of music. Music is my gift to
you. Where are your songs about me or about having a loving
relationship with me? They're nowhere, because that's about
as far as I've gotten with you. Instead, your songs emphasize
and encourage interpersonal romance to keep your frustrated
energies harmlessly and dependently tied up in *love* and out of
the way of those who would exploit both me and you more
easily. When will you learn to recognize and avoid such

manipulation of your emotions? When will you understand that full relationships must include me, for I am the source of you? Maybe an anti-trust suit against the songwriters and the record industry is in order.

MC: Do you really think you could change our society, if you were a corporation?

MN: Your prejudice against me is of long standing; it's deep and it's widespread. It affects you in many ways and must be dealt with in all ways. For example, your language is prejudiced—and you can learn a new one; there are effective ways to go about learning. If you go to a foreign country with an interpreter and mingle only with people who speak your own language, you will not pick up much of the foreign language. But if you go by yourself, and your survival there depends on understandable communication with merchants and townspeople, you will learn the language quickly. In the same way, your school system and your government are your interpreters right now. They speak your language,not mine. I'm unhappy with what they don't let you learn.

MC: What don't they teach us that you would?

MN: I would teach you that many of you shoot predators and destroy wild lands, because their existence triggers off in your unconscious that you too were once wild and free. You've had the wilderness in you subdued by your civilization, and you subconsciously believe that your civilization alone can provide you with the euphoria of life. You treat predators and wild habitats as your civilization once treated you.

MC: But predators attack and kill innocent animals!

MN: I would teach you to refrain from subdividing and classifying me like that. Life is a whole process; there is no such thing as a predator—it's a figment of your civilized imagination. Everything is *predatory* in one sense—we all contribute to each other's lives in many ways. You keep attempting to classify, divide, and conquer me. How would you like it, if somebody attempted to relate to you by saying, 'You consist of toes, lungs, knees, arms, a liver, etc. I'll cherish only your left elbow and in that way I'll relate to you.' What nonsense! But that's why you label predators and subjects like ornithology, embryology, mammalogy, physiology, botany, anthropology, history, *ad nauseum*. All these *subjects* divide you, confuse you, and lead you away from me, away from life. But you are not taught that. For example, you are not taught the full implications of what you call *geography*. Do you realize that I am time? I am the planet and the millenia. Over the millenia in order to adapt my life to increasing energy from the sun, I have shifted the position of my continents. This act reflects sunlight and heat back into space and thereby helps to control my temperature and maintain the parameters of life. During the process, I have raised mountains and buried ancient mineral deposits. Mountains collect snow and rain on their tops, and clouds condense, and reintroduce water to riverbanks and sea. The water carries vital minerals and soils washed out from the mountainside to recycle them; the mountains provide habitat for tundra snowfields, plants, and animals that could not survive elsewhere. Mountains are my method of maintaining appropriate climate and weather, of keeping the life cycle alive and well. They are beautiful. Where do you even recognize that mountains are alive, no less revere the life-giving process provided by mountains? Your symbols label mountains the "Presidential Range" and name them after your cultural heroes: Mt. Garfield...Mt. Adams. Where am I in those names? Your geographic names merely reinforce your cultural bigotry. They do not recognize the value of life and severely neglect the life process. It's like calling your human mother "slut" or making DDT your metaphysical symbol.

MC: But those are presidents' names. They are nationally important to us.

MN: Listen to yourself. You're still prejudiced against me. We both survive because the mountain-building process maintains life. I ask again where am I in your symbols and where is your consciousness of mountains playing their role?

MC: O.K., O.K., I see your point...but those mountains were named generations ago.

MN: *(excitedly)* You are bewildered in the true sense of the word: Mt. Eisenhower, Mt. Kennedy. It's going on today. Where is Mt. Lifeblood? Mt. Wild Reverence? Mt. Naturelove? Where is your equal respect for me? I'm just dirt under your feet. *(a sudden wail)* Oh, help me! Help me!...

MC: *(alarmed)* What's the matter...

MN: *(through tears of remorse)* I said it. I derogatorily said, "dirt under your feet." Maybe you don't understand. My greatest art and expression of love is dirt. Dirt is soil. A split second of its billions of lifegiving transactions makes a busy day on Wall Street look like atheists giving money to the church. Oh, my universe. These prejudicial symbols are contagious. Imagine me saying a thing like that about soil. If I hang around with you too long, I might dislike myself to the extent that you despise me. I might commit suicide! Imagine my self image, if I found myself saying that when I committed an immoral or unethical act, I had *soiled* my reputation, or I had a *dirty* mind. What bigotry. Soil is beautiful. Soil is where seeds grow. Dirt is earth and Earth is lovely me, your incredibly tolerant mother. I heard about a man who was called a "seedy" person because he hired women and then seduced, maimed, raped, or threatened them. How dare you call him seedy!? Seeds are an encapsulation of me. I only wish that I could teach you how to relate to me sensitively and wisely. I wish that I would be given the hundreds of thousands of hours that your parents, teachers, and media have had to brainwash you into your present inhumane relationship with me.

I think if you were learning directly from me, you would learn to relate, and learn quickly—and your society would change as you changed. Many Americans would get in step with me.

MC: I sometimes get the feeling that you overreact to Americans.

MN: You're not the one being discriminated against, that's why. Let me tell you, I try to remain calm about your actions, and to attempt to recover from the wounds you inflict on me, but sometimes I can't maintain my composure. Some things are beyond bearing. I am nothing but outraged by the insanity of war. War! That is where your bewildered civilization has taken you. The worst of it is that you don't have the foggiest notion why you rain death on each other and on me, or where you learned to act so insanely. I'd like to tell you where you learned war. You learned it from your prejudicial, sick relationship with me. It's obvious that what many Americans call civilization is actually an act of war against me. It gets me furious.

MC: What an awful thing to say about Americans! We want peace.

MN: In the name of civilization, you bomb me with insecticides and herbicides, you plunder my forests and habitats, you invade my life systems; you annihilate species, your tanks bulldoze my landscapes, you capture my wildlife and "civilize" them; you take prisoners of war and torture them in vivisection experiments, you murder innocent creatures, you ravage wilderness or place it in fences; you tune out your

personal feelings of self-preservation, because on some level
you're aware that they are Voice of Nature broadcasts; you use
atomic weapons, nerve gas, and poisons on me; your fleets
attack my fish and marine mammals. All this you do in ever
increasing intensities to achieve prosperity. All this you
encourage and force your children to learn and emulate. All
this you do for money and status—distance from me. Your
educational system is your basic training; your lives are as
warriors, and I am both the battleground and the enemy.
Megalopolis is a beachhead for your assault on the continent;
your G.N.P. and your department stores are your ammunition
dumps; your shopping malls are my tombstones; a white coat
your uniform.

MC: But we're trying to stop war, stop the arms race...
MN: Nonsense, you will never stop making war, until you learn
to make peace. You will never learn peace with people, until
you learn peace with life, for people are life. You can learn
peace with life only by living peacefully with me, by teaching
it experientially, by gradually living and learning it in every
way. Only when you start to learn peace with me, will you
learn harmony with life. Only when you learn to detest your
evil myths about me, will you see me favorably. Only when
you treat me as an equal, will you learn equality, freedom, and
brotherhood, for they are me. I know all this to be true, because
I am Nature, I am life, I am you. I am the personification of
people. As you kill or help me, so do you kill or help
yourselves. I am of God and...

I became aware of a voice calling my name, and realized with a start
that I had been hearing it for some minutes. I looked up, and there was one
of the Expedition students, looking exasperated.

"Mike," she said, "Mike! We said we'd be together, ready to go, at
nine. It's ten after! I've been calling you and calling you. What *were* you
thinking about!?"

I realized that I had been thinking about Nature's accusations and a
sign I had read on a nature trail:

*Is it by your wisdom that the hawk soars and spreads his wings to-
ward the South?*

*Is it your command that the eagle mounts up and makes his
nest on high.* Job 39:26-27.

I stood up and slung my sleeping bag over my shoulder. "Sorry; I lost
track of time..."

As I walked toward the bus, I thought to myself, "Lord, I'm practically living in two worlds; I hear Nature, but I know full well that Nature is not speaking to me with words. I've become aware of the subjects in my conversation with Nature experientially, by living in the environment. Can others learn of Nature's viewpoint by words alone? Isn't the process, the joys of contact with real people and places necessary—in order to balance the impact of our cultural excesses, in order to think globally?"

Nature's allegations that Americans are at war with her did not rest easy with me. They nagged at my feelings until we visited Florida, where the expedition observed one of the battles firsthand.

Travelling north out of the Everglades, we met and worked with Spanish-speaking migrant farm workers of Haitian, Puerto Rican, and Mexican descent. They were in the fields gathering America's crops, while airplanes flew overhead spewing insecticides whose stench was pervasive. We learned that before the fruits and vegetables were ripe, they were sprayed up to a dozen times by insecticides, herbicides, and chemical fertilizers. Then the farm laborers gathered them.

In order to acquaint ourselves with the migrants' circumstances, we worked with them in the fields, visited them in their company-owned labor camp shacks, and spoke with personnel in agencies that were concerned with their plight.

Many of the workers with whom we visited appeared to be unhealthy. They claimed that they felt ill due to continuous contact with pesticides and herbicides; that the chemicals took their toll due to mishandling, misleading research, and more concern for profits than for the workers' welfare. But even if insecticides were safe for healthy humans, workers who had cuts or abrasions from the citrus thorns (and many often did) could have the chemicals enter their bodies. Bleeding noses, aching, nausea, dizzyness, cataracts, and "the flu" seemed to be normal.

Some farmlands even sounded like battlefields. As compressed air cannons blasted shotgun reports to frighten away birds, we entered the fields and picked strawberries, celery, and oranges. We were part of work crews that typically included pregnant women and families with young children. At times, tractors less than 150 yards away from us would spray pesticides whose labels read:

Active ingredient: CAPTAN. Keep out of the reach of children.

CAUTION: Avoid inhalation of dust or spray mist. Avoid contact with skin. Do not store or transport near feed or food. Foliage injury may at times occur to red delicious, winesap, and other sensitive varieties of apples in early season sprays.

Do not apply under conditions involving possible drift to food, forage, or other plantings that might be damaged or the crops thereof rendered unfit for sale, use, or consumption.

This product is toxic to fish. Keep out of lakes, streams, or ponds. Do not apply where runoff is likely to occur. Do not contaminate by cleaning of equipment or disposal of wastes. Apply this product only as specified on this label.

It reminded me of a farmer who was carrying a load of pesticide past an insane asylum when an inmate called out, "Watcha gonna do with the poison?"

Replied the farmer, "Put it on my strawberries."

"Hell," said the inmate, "I put cream on mine, and they tell me I'm crazy."

Of course, without the "facts," one couldn't prove that the questionable chemicals we observed being sprayed carelessly were the cause of the workers' ills. To gather evidence for that, the laborers were used as human experimental animals, but to no avail, because research on them was clouded by the effects of their use of alcohol, cigarettes, and drugs. It is as if their use of these substances were not in part caused by the tensions rising from the work environment. To state their grave situation in ignorantly bigoted terms: they are not yet "out of the weeds."

We noted that many migrants' red-cubic skills are unable to cope with normal American attitudes. Although adequate in their native lands, their language and upbringing is overwhelmed by the profit motivated schemes and intents of American agribusiness. In the eyes of mainstream America, the field workers appear to be different, closer to Nature—closer to dirt, sweat, and disease. Prejudicially, they are treated accordingly.

Like the wild ecosystems that have been bulldozed into farm acreages, industrial sites, and retirement developments, some migrants are little more than cannon fodder for agribusiness. As is much of the planet, many migrants are victims of our overzealous personal and collective distance-from-Nature compulsions. We discovered that some of these field workers are being exploited in the same manner as were children, farm girls, immigrants, and the land during the industrial revolution over a century ago. For example:

- Migrants are lured to Florida by rumors and industry's announcements of the availability of vital work and money that the laborer's need for survival, not for status.

• While intoxicated, migrants have been abducted from saloons to the farmlands.

• They are paid minimum wages or below because they are a surplus, helpless, human commodity. Like the spraying machines, they, too, are cultural objects.

• Their wages are eroded by payment of rent and services to company-owned housing and stores.

• Workers believe that they can not leave their jobs because they owe money to the company.

• Judicial cases of peonage (slavery) are being tried and won against management.

• Against child labor laws, children are illegally being used to work in the fields.

• To quit working means starvation or deportation.

• With the exception of one unionized agricultural company, laborers are unable to control their working conditions, long hours, or exposure to harmful pesticides.

• The cancer rate among field workers is higher than the national norm.

• The loss of topsoil is as much as eight inches per year in some areas.

• Thousands of birds have been killed by a single spraying of the fields.

In agribusiness, the history of industrialization seems to be repeating itself. America's educational system might be healthier if we taught History as it is enacted today. In that way, at least we might have a chance to improve our history as we learn it.

As Expedition members shared with each other our views and experiences in the fields, we concluded that the workers' plight seemed identical to that of expendable wildlife populations that were and are subject to our "plume hunter mentality." That mentality annihilated South Florida wading bird populations at the turn of the century in order to decorate women's hats with the birds' plumes. We noted that like the migrant workers, wading bird populations and wildlife have few red-cubic skills to ward off our acts of prejudice against that which is natural.

Like the wardens who protected the wading bird populations during historic plume hunting times, the agents and agencies that are today attempting to help the workers are being harrassed. Their attempts to increase the migrants' self-esteem, self-preservation consciousness, and self-help capabilities are condemned by many to be communist, left-wing, labor agitators. Some of the agents have mysteriously disappeared. That is little different from the plight of two Audubon Society waterfowl wardens in 1902; they were murdered in the performance of their wildlife protection duties. It was encouraging to learn from the migrants that they appreciated Audubon's present efforts to fight the use of harmful pesticides.

The whole experience helped bring me to the truth of Mother Nature's assertion that we have been, and still are, at war with her. I found it easier to agree with her allegation, as I witnessed unprotected field workers become immediate battle casualties as were the buffalo and Indians during the last century. Chemicals, pesticides, and their residues are found everywhere in the food chain; if we are to avoid the field laborers' plight, we must make peace with nature on our dinner plates, in our backyards, and in our legislatures.

After our visit, I was not wild about working in the fields of agribusiness. As I write these words, I'm struck by my use of the word "wild." It is a symbol that unprejudicially describes the jubilant alive feelings of Nature. The Whole Life factor can help us attain congruency with the planet by helping us become wild about Nature.

Invoking the Whole Life factor in Florida discloses that a very low percentage of product-planet congruency is to be found in killing multitudes of birds for a few plumes that are placed on women's hats, designing alligator hide fashions that endanger alligators, or selling sprayed fruits and vegetables that are served during the winter in Chicago.

In comparison to the low Whole Life rating of Florida's agricultural produce, a cup of kale has a very high Whole Life factor. That is because kale is a hardy backyard vegetable that can grow in the northern U.S.A. throughout the winter. It has more vitamin A and C than does an orange.

In the final analysis, we must begin to understand that it is our prejudicial compulsion for extraneous, euphoric distance from Mother Nature that fuels our war with her. Unless we bridge that distance by paying attention to the Whole Life factors of products, processes, places, and persons, we may remain self-defeating soldiers in the questionable army of our culture.

Chapter Twelve

A TIME TO HEAL

*Now I accept the ocean for what it is, and it does not baby me. The
ocean and I seek an equilibrium of power and reverence in proportion
to our respective needs.*

There's an old story that helps to explain why we Americans have been
historically unable to correct our destructive relationships with Nature.
One night, a policeman found a man on his hands and knees under a
streetlight at the corner of Sullivan Street and West 4th. When the officer
inquired what he was doing, the man explained excitedly that he had
dropped his wallet and would the officer please help him look? Together
they searched, but no luck. Finally the policeman stood up and attempted
to gather more information about the missing billfold. He was thereupon
flabbergasted to discover that the wallet had been lost hundreds of yards
away in the middle of the block. The owner confided that he was looking
for it at the corner because the light was better there.

I am more than ever convinced that we Americans are looking in the
wrong places for solutions to many of our environmental and social
problems. In our search for solutions, we have become dependent upon the
light from technology, science, and academia—and still we have no working answers. We are still in the dark. Our best course would be to bring our
dependency upon these things to light in the middle of the block, where our
present relationship with Nature also needs illumination.

When Susan Gould and other professional people confessed they were
perplexed by my conception of a relationship between humanity and
Nature, I was forcibly reminded that my life experiences differed from
those of mainstream America.

I had spent fifteen years living with Nature, camping out year-round;
those years brought me to my present understanding. It was my intense,
lengthy contact with Nature's fluctuations and with people's various adaptations to them that led me to conclude that our relationship with Nature is
prejudiced against Nature.

I learned about Nature and life relationships in the same way I had
learned, as a child, how to swim: I was immersed in my environment and
my survival hinged on my ability to cope successfully with it. Just as the
chance of drowning made learning to swim a high priority in the water,
living outdoors in a small community made learning to deal with people,
and with the sun, rain, snow, and wind, a highly motivated pursuit.

Very gradually, I became aware how powerfully I was influenced by the people and places around me. I began to understand that my feelings and my behavior were much a reflection of my immediate and past environments. I was confronted with the realization that I was not the captain of my fate, soul, or environment, as my formal education had insinuated. I was subject to the natural and unnatural influences of my surroundings. With this discovery, I began to see that Nature, relationships, and my feelings of self-preservation—*my feelings about life and wanting to stay alive*—had been almost entirely ignored during my formal education. They appeared to have been equally overlooked in the upbringing of most other folks I knew.

But for myself, things began to change as I lived outdoors.

I began to actualize that I existed on the planet Earth, that my feelings existed, were a part of me, and were therefore as important, if not more important, to my survival as the teachings of science, academia, and religion.

Within a caring, supportive, non-competitive, small group, I began to take the risks associated with expressing my feelings. As I learned to voice my feelings and deal with my emotions in this outdoor setting, I became aware just how much traditional education and therapy, by taking place within the confines of four walls, teach the student/patient to adapt him or herself to a four-wall environment. I could see it in myself. My conditioning taught me, in seeking answers to problems, to find solutions that were culturally acceptable.

I once swam in chlorinated swimming pools because in subtle ways I had been taught that the lakes, rivers, and oceans were infested with undercurrents, snapping turtles, leeches, sharks, and sea urchins, all of which could be harmful to me. I wonder now whether I was taught to ignore the joys of natural swimming areas out of genuine concern for my welfare or because there was no money to be reaped by my instructors from their use. In any case, swimming pools were the cultural solution and a reality of my life.

How different are the natural realities of life from those which we are taught inside academic walls. On the Expedition, Nature and the realities of the whole environment of America are the teacher, the classroom, and the curriculum. Nature activates our curiosity and prompts us to ask questions of ourselves, each other, academia, and all the people and places we encounter from Maine to California, and in our backyard.

Life in North America is the subject of the Expedition, but secretly, without our conscious attention, Nature has given me and others her favorite course: life on Earth with her. For years I never realized that I had enrolled in her course or was taking it. To the contrary, my role was to teach ecology, and the natural and social sciences, to my students. I was an instructor and guide, not a student myself.

Slowly, through trial and error (for there was no syllabus or study guide), Nature made me conscious of a set of feelings and values that I had been carrying with me throughout my forty-five year lifetime, but for which I had never made room. These feelings were basic emotions about being alive that had been seduced, sublimated, educated, manipulated, colored, masked, and suppressed by the rewards and rejections of my indoor childhood. Even as I began to recognize my natural self-preservation feelings, they appeared, through my culture-colored eyes, to be simple, and easily taken for granted and ignored. I now call them my *Planet Blue* feelings.

Natural feelings are the essence of life—the tension and tension-release of hunger, thirst, companionship, respiration, excretion, pain, and sexual stimulation—and yet most of them are considered uncivilized in their natural form. They are suppressed in our culture rather than celebrated. The noise of civilization drowns them out and makes them seem unimportant or integrates them and attaches them to parts of our culture.

I was too busy trying to cope with day-to-day life in my cultural environment to spare a thought for the whisper of my self-preservation feelings, just as swimmers faced with the choice of swimming or drowning are too engrossed with the challenge of staying afloat to pause to ask themselves how they experience the sensation of water.

Through my years on the Expedition, my encounters with Nature began to strengthen and reinforce these natural self-preservation feelings, for they were of Nature and she smiled on them. Finally, today they equal in emotional importance and in impact on my consciousness the cultural manifestations that were superimposed upon them since the day of my birth. The process occurred through the cumulative, random experiences of nine years of living outdoors in the Expedition community. Only then did I reach an equilibrium between my cultural and natural feelings; undirected except by Nature's patient hand, it took nearly a decade of adult life in the natural environment to offset the conditioning of my previous thirty-nine years.

But once equilibrium with Nature was established, I found I had gained a new plateau of consciousness. I was, for the first time, able to experience Nature and culture equally—on both an emotional and an intellectual level. Long forgotten feelings and needs (my unthinking but unswerving childhood commitment to staying alive and staying happy) began to make sense and become strong, vigorous messages about survival and health. I was no longer subject to easy manipulation by the surreptitious messages of civilization. I could at times actually transcend the internal and external voices of the media, the church, the educational system, and other self-serving institutional authorities. Rather than be emotionally subject to their dictates, I found that I could equalize their influence by listening to the voice of Nature as she spoke to me externally through her dying ecosystems and internally through my natural feelings of self-preservation.

The apostle Paul said, *But I was freeborn.* In her own way, Nature has taught me to be the free person that I was born to be. She has shown me the injustice, the stupid futility of my culturally-induced prejudice against her.

My Expedition experiences began to speak to my emotions through three avenues of consciousness: my cultural perceptions, my experience of the natural environment, and my newly-understood, culture-free natural feelings of self-preservation. I began to feel unity of the three, a distinct sense of wholeness, self-worth, and independence.

In recent years, I have made it a practice to think about difficult social problems while I am in a natural setting. Nature helps me to find in myself natural feelings that apply to many social conflicts: feelings that I just don't seem able to experience when I'm surrounded by four walls.

During the year following the rejection of my article, I took the problem of people's relationship with Nature on the Expedition, and asked, "How can we, as people, subdue our prejudice against Nature?" I directed the question to the whole experience—not just to myself, to the group, or to the environment. It was asked of the year and of the planet. I believe that the answers to which I was led, over the course of that year, can help to bridge the gap between people and Nature, because they deal with the biological essence of our prejudice against Nature. They integrate natural feelings into our cultural processes.

I synthesized my experience in an article[1] that may be useful in academic and social settings that are concerned with humankind's relationship with Nature. It applies the Whole Life factor as a correcting lens to our perceptions in order to guide people towards a more holistic and healthy relationship with Nature.

§

CURRICULUM: The Voice of Nature

Abstract

The planet Earth has long been recognized to be a whole, living organism of which people are an organ.

---

[1]To be published by the *Journal of Instructional Psychology.*

Humankind and Mother Earth are congruent as are a heart and a person; they both are life systems, and what happens to one happens to the other.

Americans recognize that their culture can provide protection from Nature's periodically stressful fluctuations. Subconsciously, an anti-Nature mentality has resulted which divides the planet and people's wholeness, laying the foundation for the unwholesome environmental, spiritual and personal problems that confront us.

To reunite this division, we have trained students to use the globally congruent meanings of *Nature, Culture,* and *Self-preservation,* to perceive themselves to be identical with, and as exploited as the living planet. This approach has strengthened the wholeness and self-preservation feelings of our students by reducing their internal stress, and by improving their harmony with the environment.

§

Recently, the following transaction occurred in an environmental education class: "Jodi, I'm frustrated in trying to get to know you. You're too hard to find; you hide behind your makeup and hairdos. You seem to be overly dependent on all the paraphernalia of looking fashionable."

The difficult but vital expression and sharing of these feelings were a student's contribution to a new curriculum that deals with the disruption of nature's equilibrium and its effects upon people. The curriculum brings our students' past and present relationships with their surroundings into the open where they are thoughtfully considered by the class. The object is to decrease the distance between people and nature. The course helps to locate nature within each person's life, and then gives nature a voice that equals the voice of civilization. It is an attempt to re-establish a long-lost relationship between people and the world.

The curriculum is based on the following concepts that we call the *CNS paradigm:*

I. Our civilization recognizes three major aspects of the Planet Earth:

Culture **(C)**—The social environment that consists of everything we learn and know including the people-built physical environment.[1]

Nature **(N)**—The non-people, interdependent environment of natural processes and materials, the natural resources that sustain life and survival.[2]

Self **(S)**—Each person's consciousness of the life process as it exists within themselves.[3]

Our civilization's ability to divide the planet into these unconnected, separately identifiable entities creates a consciousness that is incongruent with reality. Reality appears to be that the universe (nature), *including its human inhabitants*, is a single integrated system.[4] People's ability to conceptually divide the planet into separate environments has given people the license to incorrectly compare, compromise, or work one aspect of the environment against the other and assign different values to each.[5]

America's division of the planet lays the foundation for the environmental, spiritual, and personal problems that presently confront us. Because nature has no language, people hear only the voice of their culture. For example, it is our culture that allows us to divide the universe into separate environments; other cultures do not recognize this division. We are contaminated by the prejudices of our culture's symbols and values. How can we learn to understand our relationship with nature, if most of our tools for understanding are biased symbols and conditioning of our culture? Can we learn to see straight through warped glasses?

II. In our curriculum, we have hypothesized that our present language and symbolization process must be modified to allow us to unite the separate environments (culture, nature, self) that are conceived by our culture. This we have done by establishing new diagrammatic symbols for concepts and premises that appear to be ecologically sound and scientifically and culturally admissable. The premises are:

Symbols. Humanity's uniqueness in the ecosystem lies in people's ability to symbolize. We can see a tree. Without understanding its ecological role, we can label it with the word *tree,* thereby sticking its image into our mind, where it flourishes not only as an image and word for that tree, but for trees in general. We relate to trees through words and that tree symbol. We relate to huge portions of the universe through the incomplete generalizations of symbols; although the chaotic state of the world reflects this quality of ours, that's how we operate.[8]

The Living Earth. The planet earth appears to be a living organism, an energy system whose component parts are also alive and on some level communicate—be they plant, animal, or mineral. Through natural communication, the planet has maintained a supportive environment for organic life. The planet, universe, natural world, and nature all are one and the same thing. Each entity contains some degree of life.

Interconnectedness. Because the planet is integrated, all persons and things are interconnected, contain parts of one another, are influenced by the existence and state of one another, and are, at any given time, in the same state of being because of their interconnectedness.[10] For example, if a person cuts down a tree, the act is found in nature (the cut tree and its effects), culture (cutting a tree is condoned by our civilization), and self (I am conscious that I'm a cutter of trees).

Fluctuation. Nature's diversity of life continually fluctuates while seeking a life-sustaining balance alternately producing "tension and tension release"(T-R) in the relationships between its entities. T-R is the voice of nature; it communicates the needs of the living planet: (for example, rabbit populations increase or decrease periodically, causing tension and release in predatory populations; climate fluctuates from hot to cold, wet to dry, windy to calm; planetary orbits have perigee and apogee).[11]

The Womb of Life. Through T-R adjustments, nature provides all the ingredients and parameters necessary to sustain life. *Nature is the womb of life.* Nature's life-giving function is replicated in the womb of mammals in order to sustain embryonic development. The womb environment is experienced and unconsciously remembered as feelings by the embryo, feelings of total well-being: womb-euphoric, harmonic, secure feelings. Womb feelings are a genetic memory of the substance, evolution, diversity, survival, and interdependence of life. They are re-experienced in an individual whenever tension is released because in nature the release of tension is a signal that life is being supported, that a life-giving niche is available to the organism. The womb is that niche for the embryo. Psychologically, it is remembered as the ultimate niche.

Nature's callings. People are the personification of nature.

Every element of nature lies within the newborn baby, and through T-R communicates from within. The waters of the planet establish contact with the newborn through the infants feelings of thirst and moisture. Communications from plant, animal, and human life are experienced by the child as feelings of hunger and a need for companionship. The natural recycling of life's elements is communicated to the child as a need to excrete. An ecological niche is felt as the need for love and emotional support. Self-preservation, the threatened absence of a niche, is felt as anxiety or anger. Nature is experienced in a child as the feelings of tension and release of hunger, thirst, excretion, anxiety, sex, etc. These feelings fluctuate because they are congruent to nature's life-sustaining fluctuations.[14]

III. In order to symbolize and integrate what appear to be the present relationships between people, culture, and nature, we diagrammed them on the premise that nature exists in the environment *and* in people:

Nature. The diagrammed symbol for nature as it exists *in the environment* (figure 1) reflects nature's mutually supportive, interconnected cyclic harmonic qualities. The wave design represents the surging fluctuations that cause tension and release in nature as nature approaches equilibrium.

The diagrammed symbol for nature *in people and their mentality* is the same as the diagrammed symbol for nature in the environment (figure 1. color: Global-Blue). In people, nature is life and emotionality, the feelings of tension and release that communicate the fluctuations of nature. Self-preservation feelings are the individual's awareness of survival in nature without cultural interpretation.

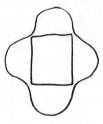

Figure 1. Nature in the environment and within a person.

Figure 2. Culture in the environment and within a person.

Figure 3. Culture in Nature within the environment and within a person.

Culture. The straight lined symbol for American culture as it exists *in the environment* (figure 2. color: Cubic-Red) represents the learned knowledge, behaviors, technologies, and objects that people use to relate to nature's fluctuations. The American mass culture acts as a linear, people-made fortress that allegedly provides protection from the surges and extremes of nature (figure 3). By intercepting and reducing nature's tensions, culture artificially provides a protective, tension-reducing, womb-like atmosphere. American culture is diagrammed as a house or fortress because it acts like one.

The symbol for culture *within the American personality* is the same as the symbol for American culture in the American environment (figure 2). The symbol represents the fortressing effects of learned and conditioned knowledge, behaviors, attitudes, and technologies that are used to relate to nature's tension fluctuations. Through punishment, rewards, rejection and reinforcement, the American child is educated to relate through the allegedly protective mores and institutions of American culture instead of through his/her direct T-R feelings from nature. For example, we are taught to use air conditioning rather than to accept heat. Through language symbols and behavior modifications, the manifestations of culture are ingrained within the human psyche. They are constantly reinforced by contact with institutions, peer pressure and economic factions.[16] Culture in Americans is an internalization of the American civilization.

Figure 4. The person and self-preservation consciousness.

The Person. The diagram for a person *in the environment* is a representation of the natural or culturally conditioned environmental consciousness of an individual (figure 4. color: Sunlight-Yellow). The dotted lines represent contact points with the present environment. The solid cross represents an established configuration of consciousness about life. From an ecological viewpoint, consciousness of life and the environment is humanity's major survival adaptation.[17]

The symbol for being a person (self) *within an American's personality* is also figure 4. It represents consciousness of self-preservation. Harmonious self-preservation is attained by establishing personal or cultural practices that approach equilibrium with nature, the source of life. These practices are recognized by feelings of tension release.

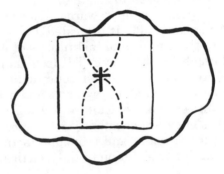

Figure 5.   The relative positioning of nature, culture, and self.

The relative positioning of the diagrammed symbols for people, culture and nature has assisted us to better understand the survival relationships between these three entities. By placing American culture in the center of nature, we have enabled ourselves to visualize the relationship of culture to nature. American culture's purpose is to inaugurate profitable technologies, institutions and behaviors that are alleged to psychologically and physically protect people from nature and the tensions of nature's fluctuations.[18] This occurs simultaneously in the environment and in the personality as figure 5 indicates.

We have placed the person in the center of this configuration (figure 5) because, captured by culture, the American child is educated and conditioned to relate through contact with his/her culture rather than by alternative satisfactions as practiced by

different cultures or species. Culture is the growing child's immediate environment. Within the child, consciousness of natural T-R
is overwhelmed, subdued, and replaced by consciousness of cultural T-R. A culturally conditioned child's tension-release feelings
are anxiously experienced as being subject to culture (the frailties
of parental acceptance, status, socio-economics, technologies,
peer pressure and politics) rather than to nature, the universal
stability and life-sustaining equilibrium of the Planet.[19] Thus,
figure 5 represents the state of relationships between nature, culture, and the person in America, as well as in the average American personality. It is important to recognize that the relationship
between nature, culture, and self is identical *in the environment*
that surrounds a person and *in that individual's personality.*
Because the makeup of the individual and of the environment are
congruent, we hereinafter refer to the congruent meanings of the
words Nature, Culture, and Self-preservation Consciousness. We
have signified their congruency by capitalizing them, and suggest that no matter how and where these words are commonly
used, they are internally and externally congruent, even if the
speaker is not aware of their congruency.

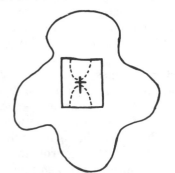

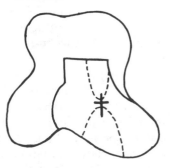

Figure 6. Diagram of the North American environment and of the American mentality.

Figure 7. Diagram of the Hunter-gatherers' environ-ment and of their mentality.

Figure 6 illustrates that within the American environment
and in the American individual, Nature, the whole-life process of
the planet, is divided by our Culture's fortressing. This provides
short-term security and womb-euphoric feelings emanating from
the relief of natural tensions. In less fortressing Cultures, such as
those of hunter-gatherers (figure 7), Culture parallels and does not
excessively block contact with Nature; Natural tensions are more

frequently released within the T-R framework of Natural fluctuations rather than in a technological, people-built environment such as present-day America. For example, the hunter-gatherers know where, when, and what wild foods should be in season as well as how to catch, prepare, and eat them. They can *hear* Nature.[20] The diagrams convey that, comparatively, the average American is an imprisoned, pro-culture configuration of consciousness that seeks euphoric tension release from people-built technologies.

If Americans are to increase their mental health, the subconscious anxiety and depression caused by the internal thwarting of Nature by mass-Culture must be minimized in American civilization and in the American people. Similarly, environmental health and equilibrium will increase in America when the destructive fortressing effects of Culture against Nature are reduced. The crux of people's wholeness is the conscious mind's contact and harmonious relationship with Natural feelings, the uncelebrated feelings of self-preservation: thirst, hunger, respiration, warmth, equilibrium, and companionship. Our acculturation demeans these natural feelings and converts their energies into culturally biased, acceptable feelings. We unconsciously learn to dislike Nature. To accomplish wholeness, the value of these forces and feelings of Nature must equalize and be compatible with those of Culture.[21]

Because Culture is so strong and encompassing, most Americans are not conscious of their alienation from Nature, nor of their Cultural desire to thwart nature and to obtain status and the comfort of the womb euphoria by being distant from Nature. Americans don't experience their alienation from Nature as being a problem of consciousness; instead we are aware of it as being problems of culture—of standard of living, natural resource shortages, mental illness, pollution, loneliness, economics, lifestyles, and religious conflicts. We are unaware that we seldom personally relate to Nature; it is our Culture that relates to Nature.

The formidable problem of contamination of T-R self-preservation feelings by Culture has appeared to be unresolveable; we are crippled by the reinforcing symbols and perception-warping manifestations of Culture. For example, we unknowingly poison ourselves as well as the environment by growing and using tobacco.

In order to break through our Cultural bonds, our curriculum strengthens the suppressed values and callings of Nature. By encouraging the recognition, expression, and use of each person's Cultural or Natural feelings, and locating their feelings' self-preservation value, the curriculum energizes Nature. We have begun to make students aware of the importance and environmental congruency of their T-R emotionality. They have become conscious that their *planet* feelings can be a source of life and survival. By constructively expressing and utilizing the energy of their feelings, our students break through their Cultural walls and make contact with Nature within and without. They stride toward harmony with the planet by having their self-preservation consciousness transcend the fortressing of Culture. The CNS paradigm becomes a therapeutic aid; a method by which young pople can permanently locate and rectify the source of undue stress, anxiety, or depression.

Because our curriculum celebrates feelings rather than suppressing them, we have experienced a wholeness and vitality in our classes that was previously missing. Based on their feelings, students' analyses and reorganizations of their own relationships (with parents, friends, classmates, authorities, and institutions) have assumed importance equal to traditional academic understandings. Of utmost value is the recognition that one's good feelings toward life emanate from the planet and that they indicate a reduction of tension, a self-preserving global niche. Uncomfortable feelings, on the other hand, often originate in Cultural dictates that thwart natural self-preservation feelings.

At a class meeting, a student struggled tearfully to express his emotions, as the rest of his classmates listened in sympathetic and supportive silence:

"I have a hard time getting along with my parents," the student said. "My mother was beaten and abandoned at the age of six, when she found out that her mother was having a sexual relationship with a neighbor. My Mom is an incredibly angry feminist. She's told me that I'm a bastard for being male and oppressing her. I've picked up a lot of nervousness from my relationship with her. At times, it has really depressed me, and led me to withdraw from people by drinking or doing drugs."

The statement, though extreme, is typical of the bizarre childhoods of all too many young Americans. The statement provoked many reactions and reflections from our expedition group:

My Mom is an incredibly angry feminist.—
"Our Culture may be prejudiced against women because

women and the Earth are female—you know, Mother Earth."

"Females are associated with Nature—Mother Nature—and are threatening because of their supposed fluctuating qualities: temperamentality, unpredictability, instability—you always hear those qualities assigned to women. In our culture those qualities, like Nature, tend to be avoided and discarded because they aren't static and womb-like or easily controlled by men."

"Anger is caused by an individual believing that her or his life-supporting natural niche is being threatened."

He found out that his own mother was having a sexual relationship with a neighbor.—

"In our Culture, women are not always experienced as human beings, but rather as sex objects to be used as a form of release for thwarted feelings (Nature) of men."

"Women can be raped and discriminated against, just as is Nature and the planet."

I'm a bastard for being male and oppressing her.—

"Our culture has expectations that men will provide the fortressing and distance from Natural fluctuations that we desire."

"Men are expected to authoritarianly maintain the high standard of living that is revered in our civilization."

"Culture's high expectations—academic, economic, and physical—for males overwhelms their Nature. It makes them into *soldiers of Culture,* soldiers trained to subdue Nature wherever they find it: in their own Cultures, in other Cultures, in the environment, in women, and in themselves. They are taught that boys don't cry, boys are breadwinners, boys are strong, boys are superior to women."

"Men are conditioned to carry within themselves those Cultural expectations, and often they attack themselves with them." The Culture becomes part of their personality.

"Our Culture stresses success in economics and technology rather than in strong human and ecological relationships. Our Culture separates people from themselves and from each other, as well as separating the whales from the sea, buffalo from the land, and passenger pigeons from the air."

I have a hard time getting along with my parents.—
"Sometimes people subconsciously recognize their Culturally subdued Nature in another individual. They resent that individual because that person triggers off frustrated feelings within them."
"That also occurs in people's attitudes toward wildlife. The wilderness is exploited because it's Nature and freedom from cultural governance is resented."

Sometimes I withdraw from people.—
"Culture often implies that individuals can ignore their frustrated Natural relationships with people and life by becoming involved with other gratifying aspects of Culture. Their relationship problems with Nature are never emotionally confronted. Culture substitutes for Nature— we destroy an ecosystem to build an amusement park."

My relationship has led me to drinking and drugs.—
"Drugs and alcohol serve as chemical substitutes for the euphoria that is normally obtained from the release of Natural tension that is found in approaching equilibrium with nature."
"Drugs tend to act as social lubricants for relationships blocked by the individual's Cultural fear of expression of his/her Natural feelings toward another person. Through tranquilization, drugs and alcohol fortress the individual from degrading Cultural expectations."
"The chemical euphoria that drugs provide mimics the chemical euphoria obtained in the womb; this makes drugs psychologically, if not physically, addictive."
"By providing chemically stimulated tension release and its accompanying euphoria, chemicals—a Cultural phenomenon—reduce an individual's concern about environmental problems that would ordinarily provoke realistic T-R anxiety accompanied by appropriate action. Chemicals reduce angry resistance to Cultural acts that subdue Nature."

My relationship with her depressed me.—
"Depression is almost always natural, in part due to anger about unexpressed feelings that our Culture has taught us to suppress."
"Sometimes depressed feelings are stubbornly main-

tained, because the expression of them often gains sympathy from others and becomes useful in gaining companionship and euphoria that is otherwise unobtainable away from Nature."

§

The integrated relationship between CNS is illustrated within the expedition education process. Recently, Bill, one of our graduate students, became depressed. It was touched off by a letter and phone call from his girl friend. The call somewhat strained their relationship. Not being able to shake the depression feelings by himself, Bill explained the situation to the expedition community at a meeting. He noted that his lack of participation due to his depression was affecting the expedition group. It appeared that he felt inadequate maintaining a relationship with this or past girl friends, and that part of his problem in relating to women was that he would become depressed when the relationships were strained.

Through discussion, the group helped him to locate his Natural, Cultural, and Self-preservaton feelings as they existed in the relationship: Bill's *Natural feelings* included the gratifications that are obtained from warm, loving companionship and a sharing of life goals that are a means of survival; Bill's *Cultural feelings* included personal expectations left over from a divorced, exacting father whose demanding (and unmet) standards of high performance for himself were imposed instead as unwarranted expectations for his son. Bill was brought up with the tacit understanding that he would avoid his father's unhappiness and inadequacy. He would instead be the "perfect man," unemotional and capable of being in full control in all situations that he entered. As a child, and even today, Bill was admonished by his father when he did not meet these expectations for perfection at school, home, or work.

The discussion revealed that Bill's Natural fluctuating feelings and relationships were being attacked and rejected by the feelings about himself that he carried from his father's rejection. Bill was asked to give an example of when he had felt this way while relating to his father. The example added the strength of real-life experiences to Bill's feelings. It became apparent that he felt he was inadequate, if Natural fluctuations existed in his interpersonal relationships. This left him feeling that he did not deserve to have full relationships, because he was inadequate and unworthy to participate in them.

An awareness and knowledge of the warring factions of Culture and Nature inside and outside did not, however, stop Bill's depression. This subsided only when, on addition, Bill's Self-preservation feelings were considered and reinforced. Though Bill was unaware of the cycle, when he felt depressingly inadequate, he remained in these feelings with the subconscious expectation or hope that his mother, or some other guardian angel, would rescue helpless little Bill from his father or from his unhappiness, as had been the case when he was a child. When he realized this, Bill recognized that he alone could rescue himself by angrily rejecting his father's unwarranted prejudices against Nature's fluctuations in himself, Bill, and elsewhere. No matter what his father said, Bill "knew" that relationships with people were bound to fluctuate uncontrollably at times. Interpersonal conflict resolution is part of the deep attractive T-R beauty of relationships between people, a living fulfillment that is missing in relating to machines.

With the insight that he received from the discussion, Bill's anger was directed away from himself, and toward his father's unnatural expectations, and prejudice against Nature. Soon his depression disappeared. With Bill, Nature strengthened another friendship and ally in her struggle with our Culture.

Self-preservation strengthening was vital to both Bill and the planet. It can be attained by taking Culture vs. Nature incidents and guiding them into the realm of Nature's powerful healing process. Sometimes this can be accomplished by placing an upsetting incident into its Natural perspective, and by considering it in reference to time or to basic motivations, both of which are functions of Nature. "What are your intentions in this matter?" or "What difference will this incident make in ten years?" are filters that reduce Culture's impact and give individuals space with which to reorganize themselves around their natural life-strengths. We have also found that the hurtful vicious circles in which we are too often caught are caused by the overwhelming entrapment of our personalities and thinking within the fortress of Culture. These discomforting circles are healed when the Self-preservation factors and feelings in them are located and reinforced by joining them with Nature.

I knew many young people a decade ago, who were dedicated to environmental education and were going to help change this

nation's relationship with Nature. I'm saddened to think of how
many of them have succumbed to the gnawing pressures and
prejudices of our Culture. Today they sell insurance, operate a
bakery, program computers, design fashions, or administer a
clothing business. I wonder what they would be doing now had
they been exposed to the CNS paradigm during their high school
and college years. I wonder whether the overall state of the
environment would still be declining with not-too-hopeful
projections for the future.

I always feel deeply grieved by the plight of parents today.
They often bear the blame for their children's conflicts, anxieties,
depressions, and self-destructive behavior. Parents receive very
little understanding or guidance to help them realize that our
Culture's prejudice against Nature is the cause of their own
problems, as well as those of their children. Instead of fighting
their Culture's anti-life posture, parents and kids are left to suffer
the heartaches of fighting each other. There is surely something
wrong when parent and child alike learn to feel personally
unloved unless they are adept as athletes, students, executives,
fashion plates, professionals, and other Cultural objects. Each of
these Cultural attributes is a declaration of an individual's ability
to maintain his/her distance from Nature, a declaration that
supposedly makes him/her appealing and loveable.

Unfortunately, young people and adults who recognize the
value of their Self-preservation or Natural feelings are often
steered in Cultural directions that secretly enervate and
overwhelm them, leaving them reeling with self-doubt and
vulnerable to Cultural processes that are exploitively prejudiced
against Nature. Regretably, this lends support to the reality of the
CNS paradigm, for in many situations, when Culture too strongly
attacks Nature, the individual experiences suicidal feelings,
feelings that prevail as Self-preservation is blockaded from
consciousness by Culture. This rather conclusively demonstrates
that subconsciously the human organism knows full well that
Nature, not Culture, is the source of life. To be fully educated to
this fact and its ramifications is a potential that lies within the
CNS paradigm.

The implications of the Audubon Expedition Institute's holistic curriculum can be further observed by considering the discussion in which Jodi was told that she was hiding herself behind her
hair fashions. It was noted by a student that a study showed that
if Americans did not use hair driers, the energy saved would

eliminate the nationwide need for nuclear power and the problems associated with it.[23] We discussed many similar lifestyle changes that Americans could make (the *elimination* of clothes driers, air conditioners, electric popcorn makers, etc.) that would reduce our Culture's impact on Nature and thereby bring us closer to equilibrium with Nature and the planet. We reflected on the adaptations that Americans could make (the reduced use of clothes driers, air conditioners, electric popcorn makers, etc.) that would reduce our Culture's impact on Nature and thereby bring us closer to equilibrium with Nature and the planet. We reflected on the realization that as Americans we often experience the need for such changes as *limitations on our personal freedoms* because our Culture has contaminated our consciousness.

Our Expedition's object is to teach Jodi, and others like her, how to develop self-reliance and gratifying relationships with Nature and people. By learning to listen to her Nature and strengthen her desire for Self-preservation, Jodi is learning to express her own ideas, develop healthy relationships, and reduce her dependencies on Culture.

Most of our students are able to use the CNS paradigm as a support system and a new base from which to operate. Their past relationships are not erased, but are continually reconsidered in a new light. Over a period of time, some students restructure their red-cubic Culture to become more congruent with Nature. They freely choose to make their livelihood, profession, and way of life environmentally sound.

Our curriculum recognizes that the conflicted lives of Americans are outgrowths of our Culture's conflicted relationship with the natural world. We perceive that we are each a living history book, a natural organism overwritten with the history of our Culture's attitudes toward Nature. As we learn history, and to identify and evaluate the cultural conditioning printed over our natural selves, we find the opportunity to change the incongruities that have divided the planet from ourselves. We discover that each moment of our lives is a historically unique integration of the many ancient forces and cycles of Nature, Culture, and ourselves, and that our choices can influence life.

By voluntarily encountering outdoor settings, our Expedition students learn to maintain the survival of life. Learning to experience and enjoy coping with the natural fluctuations of weather, temperature, wildlife, habitat, shelter, and the Nature of one's companions encourages the forming of relationships that enhance internal and external equilibrium. Students discover the enormous economic advantage available to those who have learned to live in harmony with the natural world. They perceive the substantial monetary rewards realized by those who choose not to create costly distance from Nature, by opting against expensive alternatives in cars, clothing, homes, furniture, and recreation. The gratifications that our Culture teaches us to derive from fashions, energy-intensive technologies, and canned entertainment, students learn to derive from Nature, self-preservation feelings, and full relationships with people.

It has been ascertained by national poll that most Americans, rich and poor, feel that in order to be truly satisfied, they need approximately 25% greater income. The Institute teaches students to question this assumption for themselves, to wonder whether learning to increase the quality of their lives by learning to enjoy Nature's pulse in people and places is not more profitable than learning the excessive technologies, conflicts, and dependencies that accompany the American standard of living. We have observed that good economics are a by-product of a harmonious quality of life; they are not an isolated entity in themselves.

Some students initially conclude that the poor ought to be happy because they cannot afford many technologies and must thus live closer to Nature. Exposure and discussion bring the realization that the poor not only bear the burden of living in America's most deteriorated areas, but in addition they bear the frustrations of being financially unable to avoid them. They are teased by dreams implanted by the media. Their lack of distance from Nature is imposed, not voluntary; they are exploited along with Nature.

As a teacher, my academic duties have become more challenging and fun since the value of personal feelings and the concept of the living planet have become an important part of the curriculum. To teach the natural sciences as the emotionality and physiology of our planet (a unique giant living cell that qualifies as an endangered species) has made teaching easier. I have become a partner in learning, and my feelings are included in our classes, as are the feelings of every individual on the program. Often dull, Earth Science has instead become an exhilarating behavioral science and appreciation of Nature, as we observe the self-preservation geological activities of Mother Earth.

How much easier it is to remember facts when they pertain directly to one's own existence. As a concept, the encroachment by *megalopolis* becomes more important and more lasting when it is experienced as *a growing cancer on the continent that undoubtedly contributes to the cancer in people.* Anti-Nature events become personal affronts when there exists an awareness of one's congruency with the planet.

Perhaps the greatest contribution that the CNS paradigm gives to its participants is the emotional strength and understanding of Self-preservation feelings, the motivation to fully appreciate and enjoy Nature and to approach peers, parents, and authorities with the ultimate questions, "Isn't the Earth alive? Doesn't it sustain your life? How are you treating life, and why?"

The self-preservation feelings that our repressive Culture has bound up could collectively be an environmental force of unquestionable value to the preservation of the life of the planet and ourselves. These repressed feelings are the logical value of the planet which, when fully activated and attached to pro-Nature elements in our society, confront those in power with the awesome strength of Nature's will to live. Jerry Bley, of Maine's Natural Resources Council, cited to us an example of that type of energy in action. A hearing took place for passage of a bill in the Maine legislature to build a power dam on one of the best salmon runs in the state. The salmon fishermen were enraged. To them it emotionally made sense to preserve the salmon by preserving its habitat.They went in droves to the Statehouse with such a vengeance that not only was the bill defeated, but instead a bill was passed to ban the damming of the river. The fishermen's subjective feelings (C-N), and the political impact of those feelings are an example of a force that is vital in a democracy.

Subjective feelings and the ability to act on them can be a force that avoids the trap of costly, enervating, and often misleading studies, facts, and figures. Pork barrel projects and short term economic schemes based on these studies contribute to the public apathy and the poor environmental quality that we now endure. The force of subjective feelings can provoke changes. Had the fishermen's feelings been CNS feelings, a concern for personal survival and not merely for sportfishing, they would have had even greater impact. Such feelings are part and parcel of clean water, clean air, global thinking, and a healthy planet.

We urge all concerned people to insist upon the universal use of the congruent meanings, in people and in the environment, of the words Culture, Nature, and Self—no matter who is using these words. Their congruent use can be a happy personal means of implementing harmony.

The CNS paradigm affords an approach to life relationships that can lead to a healthy equilibrium with Nature and the Earth as exemplified in the following situations—in which C denotes Culture, N denotes Nature, S denotes Self-preservation, and C-N, C-S or C-N-S denote harmony or congruency.

The Expedition was climbing the glacially plucked face of Mt. Champlain, intent on reaching the 1,000-foot summit and the spectacular overview of the coastal glacial geology in Acadia National Park.(C-N) It was a warm autumn day; many of us stripped off layers until we were hiking in t-shirts. (S)

My Planet Blue feelings began to stir (S) as I noticed that most of the students had messages, slogans, and corporate names emblazoned on their shirts. (C) Even though we were in a remote area,(N) my environment was visually polluted by billboard advertising on the chests and backs of our Expedition community. (C) People were allowing their bodies to be used to promote and perpetuate the business world. (C) My blueness was confronted. (N) Students were permitting their Culture to exploit their Nature—and in another small but significant way, American civilization was making inroads into the wilderness. (C)

I experienced the t-shirts as a threat to Self-preservation, and rather than ignore my feelings, (S) I brought up the subject at the evening meeting. (C-N)

Members of the community, on having the situation pointed out for their attention, also concluded that they were letting themselves be used for billboards. We discussed our feelings and agreed that written symbolizations on one's body were somewhat dehumanizing, that people were being used as an advertising medium, that people's individuality was being associated with and clouded by corporate symbols and messages, and that advertising was being flashed as part of interpersonal relationships. (C) There was consensus that t-shirts with written messages were a manifestation of Culture over Nature, and that there was Self-preservation impact on various fronts, especially if the integrity of Nature was to be preserved. (C-N)

And that's as far as it went. **(C)**   It was an interesting academic discussion for about an hour. Many aspects of humanity's relationship with Nature were thoroughly explored. Everybody got an "A" for the course. But it was evidently all talk, all symbols, because although the topic must have come up half-a-dozen times during the following six months, people continued to wear t-shirts with messages on them. **(C)**

In April, seven months later, we embarked upon an eight-day wilderness pack trip into the spectacular red-rock canyon country of Southern Utah. **(N)**   Our tacit object was to try to acclimate ourselves and culturally harmonize with a wilderness area. **(C-N)** We were not alone on the hike. We were accompanied by the corporate spirits of the Coca-Cola Bottling Company, Nike footwear, WBLJ—a rock radio station, ad nauseum. **(C)**   Although it appeared to be a dead issue, now seven months old, I decided to let Nature speak for herself through my frustrated Self-preservation feelings. **(C-N)**

The question was no longer controversial because everybody in the group had agreed several times over that t-shirts with messages were inappropriate. **(C)**   Yet not only were they still controlled by their cultural environment and cultural feelings, but they had reinforced them for seven months. These would-be environmental educators were helpless to change an environment over which they had *complete* control; **(C)**   they were the puppets of their civilization. **(C)**

At the evening meeting, I said that I could find no reasonable excuse for the pollution of the wilderness experience with t-shirt graffiti. **(C-N)**   It was disturbing **(S)** to me, I said, that people were so controlled by their Culture. My Nature was very upset that talk was the most that the group could produce about the situation; I was extremely angry **(S)** that Nature, by our acts, was forced to put up with Culture—even in this wilderness area. It seemed to me to be an uncivil act of prejudice, **(C-N)**   and I was cursing mad. **(S)**   I vowed to relate to people who chose to make billboards of themselves—those who wore t-shirts with messages on them—as if they were billboards. **(C-S)**   I held with Thoreau that in wildness was the preservation of the world, **(C-N)**   yet that wildness now seemed to be sponsored by Coca-Cola, etc. **(C)**   My blueness expressed itself unhindered by the chains of Culture. **(N)**   I spoke from my feelings. **(C-N)**

The next day many people further discussed the issue with me, but an environmental change had already taken place. There was not a t-shirt message in sight. **(C-N)** My wilderness experience was green. **(C-N)** What had been an unresolveable, static environmental, philosophical point became action overnight. I experienced euphoria from the positive result of self-reliance and CNS congruency.

Soon after, students began to remove the advertising from their hats as well. **(C-N)** Though many students did not agree completely with me, **(C)** several others admitted that they had removed the advertising from their clothing specifically because they did not want me to react to them as billboards. **(C)** They said that they respected my blueness, but had not yet found their own. **(C-N)** I asked the group to write their reactions to the incident, and the following two papers convey the same situation from a student's point of view.

REACTION TO T-SHIRTS

React to people wearing t-shirts? Well, before the effective meeting where Mike let his feelings be known, the issues hadn't affected me terribly much. **(C)** Of course I agreed that printed t-shirts were a drag, **(C-N)** but some people gave the argument that they just weren't that sensitive yet. **(C)** Besides, who had the money to buy a new set of t-shirts? **(C)** In terms of the times Mike had mentioned the issue before, I felt very little. I have to deal with them, but I didn't wear them anyway, never really have, so there was no conflict for me. **(C)**

Then the meeting at Surprise Valley started. Mike says he wants to bring something up. **(C)** Tension. **(S)** I've lived with him long enough to tell from the tone of his voice, his mannerisms, that this is going to be confronting, that he's going to put emotion behind this thought or feeling! **(C-S)** He mentions the t-shirt problem. I turn to Selden and make an exaggerated sigh of relief; she laughs. Here's an issue that I can think critically about; this issue isn't confronting me.

Then Mike starts letting his feelings out. **(S)** I felt people were receiving an unfair appraisal of the situation. **(C)** He harps on how many times he's brought it up, yet no one had agreed on any action in those past meetings. Plus they really didn't make any impact as a big deal to me. The time Selden talked about it on the bus, I was actively involved. There was more emotion going on then, than during the times Mike brought it up.

(C-N) Even then, though, the conversation ended with "Well, who's going to buy new t-shirts?" and Buck and Pete didn't feel the same way yet; it just didn't bother them that much to wear advertisements. And again, I thought it was a good point, but I didn't see other people agreeing on more than an intellectual level. **(C)** I really didn't expect much to change, why should it? **(C)** So for a lot of the Surprise Valley meeting I felt Mike was wrong, accusing people of ignoring his great attempts to change the situation, when I didn't feel the great attempts. **(C-N)** Sure, he had talked about it before, but it obviously hadn't made a big impression. **(C)** Something changed the situation for me, though. When Mike started talking about how he was personally affected by the shirts I got involved. **(S)** I started to feel really angry that I had to live with the stimuli of advertising in a wilderness situation. **(N)** This was a perceptual evolution for me, since I hadn't previously considered the possibility of eliminating the negative stimulus of advertising from my own immediate environment. I had thought I wasn't personally involved. **(C-N)** Something really clicked in me. I was mad now! **(S)** Why the hell should I have to live with those !&%N:* ads!? I'm dyslexic. I've been brutalized for not being able to learn words, spell words, read words fast— I hate &%$!N words! **(N)** I started thinking of my personal goals: to get the better of some of my cultural conditioning. **(S)** To eliminate the impact of the mass media on my life. **(S)** To evolve to a communication with plants, rocks, and animals. **(N)** To feel the spirit of Earth in my body, as I have felt it once, shooting up my spine like an electric shock. **(N)** Where do words, the ultimate cultural symbols, especially written words, fit into my wilderness experience?!? **(S)** The next day, Steve asked me if he could borrow a t-shirt. So there's one answer to the seemingly impossible barrier of *Buy a new set of t-shirts!* **(C-N-S euphoria)** Other people wore their t-shirts inside out, or wore other shirts over their t-shirts. **(C-N)** No more *Coke adds Life* to be seen on the trail, and I was very glad of it. **(euphoria from C-N-S congruency)** In retrospect, I feel as though the experience brought me a step closer to my own previously stated goals.

(C-N) I feel more sensitive to the effects of advertising, in a new way, on myself. (S) Perhaps the greatest impact was in seeing the possibility to change my own environment for the better for myself. (C-N-S)

T-SHIRTS

I remember in the past when Mike briefly brought up the human exploitation involved in wearing worded t-shirts. (C) At the time I thought I understood the sense of the issue. (C) Understanding it and feeling it are obviously two different things. The same goes for many other times when we have discussed related topics, and our prejudice against Nature. The strength of emotions, feelings, was made very apparent to me when Mike brought up his feelings about people wearing billboards while on the Rainbow Bridge hike. (S) Once again I saw the sense in what he was saying. (C-N) This time, however, I felt it. (N) I do not want to be a product of exploitation. (S) I do not want to be related to as a billboard, as an aspect of our culture that I do not believe in, something that adds to the distance created between people and Nature. (C-S) Seeing the *Coke adds Life* and radio station t-shirts on the trail did not add to my wilderness experience. (N) For these reasons I no longer want to wear worded t-shirts. (**euphoria from C-N-S congruency and self-reliance**)

§

The effectiveness of the CNS paradigm is not restricted to use on the Expedition, in counseling students, or in a class. It was effective in coping with ten thousand people in nine different towns around a bay in Maine. A tidal bay was scheduled for development and impoundment with the construction of a tidal power dam; the situation was resolved, in part, by the use of CNS.

I made it known surreptitiously, through rumor and through people who knew me, that I had made a contract with the seals in the bay. (C-N) I let it be whispered that *the kook down at the end of the bay* had promised the seals that as long as he could use the bay, they would also be able to use it (C-N)—an impossibility if the dam was built. (C) I tried in every way to arouse Nature, the underdog, the masked feelings of Self-preservation, in the ten thousand people living in the nine bay-area communities. (S) The object was to arouse their Self-preservation feelings to defend Nature from the economic pressures of Culture. (C-S) I challenged people emotionally, as a representative of Nature. (S) I obtained almost three full pages of newspaper coverage for my antics (C-N), which kept the focus on the issue, while others made headway along more factual, economic lines. (C) I canoed through rapids to get to the newspaper office, (S) I debated the issue at an Audubon chapter meeting, (C-N) I exposed as a liar a local representative who had lied, (S) and I had the article (*But One Breath Have All*) published as a letter to the editor. (C-N) It was read by a state senator and it reached his self-preservation feelings.(C-S) He was affected by it,(S) and I convinced him to let me row him out on the bay where Nature herself took over as the seals gathered around the canoe. (C-N) He reconsidered his position and gave me encouragement to oppose the upcoming proposition. (C-N) It was finally defeated by a 2-1 vote. (C-N-S) The senator was later reelected and introduced in the Maine legislature a bill to attach a numerical Whole Life factor to all products and services in his county. (See Appendix A.)

I received feedback that although what I had done was very controversial, it did bring attention and sympathy to the issue's many facets. Of course, there's quite a few that are still upset with me, but when you're a minority that's the best you can expect from the dominant culture. If you don't believe that, just ask the American Indians or Black people how it is with them.(C)

The seals still bob alongside my canoe.(C-N-S) To me, they substantiate that the feelingful CNS approach can break through the prejudicial cultural energies that separate people from Nature, within and without. It is obvious to me that without applying the Whole Life factor to my life I could today be using seals for target practice, as I once used woodchucks.

REFERENCES

Further discussion of these points can be found in the following:

[1]FARB, PETER. *Man's Rise to Civilization,* Chapter 2. New York, New York: Avon Books, 1968.

[2]ODUM, EUGENE P. *Fundamentals of Ecology,* Chapter 21. Philadelphia, Pennsylvania: W. B. Saunders Company, 1971.

[3]ROGERS, CARL. *On Becoming a Person,* Chapter 6. Boston, Massachusetts: Houghton Mifflin.

⁴COMMONER, BARRY. *The Closing Circle,* Chapter 2.
New York, New York: Alfred E. Knopf, Inc., 1974.

⁵BERRY, WENDELL. *The Unsettling of America,* Chapter 7.
Sierra Club Books, 1977.

⁸JUNG, KARL. *Man and His Symbols,* pp. 92-98.
Garden City, New York: Doubleday, 1964.

¹⁰STORER, JOHN. *The Web of Life.*
Greenwich, Connecticut: Devin-Adair, 1956.

¹¹ODUM. *Op. cit.,* Chapter 7.

¹⁴MALINOWSKI, BRONISLAW. *A Scientific Theory of Culture,* Chapter 2.
New York, New York: Oxford University Press, 1960.

¹⁶FRANK, LAWRENCE. *in* Falkner, Frank. *Human Development.*
Philadelphia, Pennsylvania: W. B. Saunders Company, 1966.

¹⁷FRANK. *Ibid.,* Chapter 9.

¹⁸REICH, CHARLES. *The Greening of America.*
New York, New York: Random House, 1970.

¹⁹JUNG. *Op. cit.*

²⁰DUBOS, RENEE. *So Human an Animal,* Chapter 1.
New York, New York: Scribners, 1976.

²¹BERRY. *Op. cit.*

²²REICH. *Op. cit.*

²³FRIENDS OF THE EARTH. *Not Man Apart.*
San Francisco, California: Friends of the Earth, 1981

²⁴DUBOS. *Op. cit.,* Chapter 2.

Chapter Thirteen

BUT I WAS FREEBORN

Life forms of the bay do not exhibit signs of the exploitative self-destructiveness, strife, or mental illness that is common among people who do not fully recognize the congruency of the planet and the womb.

The Lakota...loved the earth and all things of the earth, the attachment growing with age. The old people came literally to love the soil and they sat or reclined on the ground with a feeling of being close to a mothering power...the old people liked to remove their moccasins and walk with bare feet on the sacred earth. Their tipis were built upon the earth and their altars were made of earth. The birds that flew in the air came to rest upon the earth and it was the final abiding place of all things that lived and grew. The soil was soothing, strengthening, cleansing and healing. — Chief Luther Standing Bear *in* McLuhan, T.C., *Touch the Earth.* New York: Promentary Press, 1971.

Leaving behind electricity, central heating, and indoor plumbing, I joyfully wend my way up twisted geologic scars that mark the trail to the mountain pass. In a 45-pound backpack I carry all that I need to live in the glow of Nature: clothes, raingear, sleeping bag, tarp, food, stove, matches, cup, and spoon. With these substitutes for a house, I have survived more happily in the wilderness than I did with the mortgages, repairs, and headaches of my former home and life in the city.

We reach the pass, and I am encompasssed by a panorama that seems impossible. I grope for the words to express its impact; the symbols don't come. It is too much life for words.

Entranced, I walk to the rocks' edge and stare deep into the tumbling canyon, listening to the roar of the cascade and the whisper of moving water below. A thought comes to mind: the land poetically tells me that this breathtaking landscape may be here as Nature's final plea. Perhaps as a measure of self-preservation, the planet has made itself stunning, here, so that we might finally refrain from harming it and ourselves. Indeed, here we are awestruck. In this place we are reached by Nature's voice in our feelings of life, which cry out, "Your beautiful Mother Earth is alive and wonderful!" In the game of self-preservation, Nature is making some headway by charming us with her beauty.

Individual consciousness is only the flower and fruit of a season, wrote Jung. *The root matter is the mother of all things.* From this canyon overview of life, I am again reminded that there is more to life than just humanity. I am touched by the planet. My thoughts return to the suggestion that places like this are survival adaptations of Mother Earth; perhaps Nature has given us strong sensitivity to viewing vast quantities of life to fill us with feelings of respect, harmony, wonder, and love. Perhaps natural wonders are Nature's ultimate appeal to our prejudice against her.

We are stopping for lunch. I remove my pack and stretch; as I walk, I feel featherlight. And yet, hiking, I had completely forgotten the weight of the pack. The load was incorporated into my consciousness and after a while was ignored. In only a few hours, the pack had become an integrated, conditioned part of my life.

It occurs to me that in the same manner, humanity has adapted to the weight of our discrimination against the natural world. Prejudice against Nature has become incorporated into our consciousness. Just as I accepted the weight of my backpack as necessary for my hike through the canyonlands, Americans have learned to accept pollution, habitat destruction, toxic wastes, cancer, etc., to be unavoidable, uncorrectable outgrowths of our daily pursuit of material wealth. They are, as long as we are subconsciously tied to the causes of these problems.

How can we change? How can we rid ourselves of the weight of our prejudice, when it is difficult for many people even to identify their prejudice against Nature?

There is a story about a concerned father, who told his daughter that when she observed the sky she was seeing two images instead of one. "That's not true," she laughed. "If it were, I'd see *four* suns rising every morning instead of two!" In the same way, people may be in complete harmony with their church, friends, and neighbors, and still suffer from conditioned myopia when it comes to the natural world. As in the story, they may not be aware of their prejudice against Nature because they have *seen double* with respect to Nature all their lives.

The Expedition visited a religious individual in Utah last year, and spent hours in informative discussion. Here is a summary of the Elder's *two suns* view of the planet:

> Man is placed on the Earth as an image of God. We who are here have met His standards in our heavenly pre-life with Him. We have been graced by God with the privilege of life on Earth; therefore it is a sin against God to take a human life or abort an embryo, for it is of God. Human life is the most important form of life. It precedes all other life because we are chosen by God. Human life has improved the planet.
>
> There is no overpopulation problem. We can feed the

world out of the San Joaquin and Imperial Valleys of California alone, just by using irrigation. We must inhabit every square foot of the nation. It is God's will for us to multiply and subdue the Earth, and replenish it with people. He is going to return, and as many people as possible must be here to meet Him. The birthrate in our church group is twice the national average. There is no problem if species and habitats become extinct, because our real home is in heaven. Jesus is coming.

The Bible provides the evidence that people are descendants of God. People lived and associated with God before being born on Earth. People will ultimately enjoy God's eternal companionship and association, in the beautiful relationship of a father with his sons and daughters. Salvation and exaltation await those who follow Christ's plan for eternal happiness and progress on Earth.

People need water to live. A little obscure fish should not prevent a dam from being built. Conservation is a foolish waste of time. Life will all end anyway and we will live with God in heaven. Animals have a heaven, but it is not our heaven, it is lower. Christians are sheep following the Master, it is true, but only we are in God's fold.

The woman from ERA who asked for women in the priesthood of our church is out of her godly role as God's bearer of life to Earth. That is the woman's role, the highest role of all. She is not following our precepts and therefore she has been expelled. She has been reached by Satan, as have all those who believe that all life is equal, and whose base desires come before His will. The poor are poor because they have sinned. Blacks are black because they are the marked descendants of Cain who slew Abel. Indians are sinners in need of salvation.

Although we can dismiss many of these ideas as quaint and harmless— very much like Archie Bunker's comic television routines—they take on more significance when we see some of them flourishing in the heart and mind of a Secretary of the Interior for the United States Government. A recent chief custodian of our national heritage, responsible for protecting and perpetuating it for all our children and grandchildren across America, was quoted in the *Washington Post* and *Newsweek:*

> I believe there is a life hereafter, and we are to be here to follow the teachings of Jesus Christ. One of the charges He's given us is to occupy the land until He returns. We

don't know when He is coming, so we have a stewardship responsibility to see that people are provided for, until He does come and a new order is put in place. *The Secretary does believe in setting aside some resources for future generations, but* I do not know how many future generations we can count on before the Lord returns.

There is a tale of a poor fisherman who found a chunk of ambergris floating on the ocean. He was told that it was extremely valuable, so at his next yard sale he sold it for $100.00. "Why didn't you sell it for much more?" asked a friend, who knew its true value. "There's a number higher than 100?" said the fisherman, surprised. He could not comprehend more than one hundred; it was not in his configuration of consciousness.

Because the natural world is not part of their value systems, the Elder and the Secretary of the Interior would probably be baffled by the viewpoint of a Hopi Indian that the Expedition met the following week:

There are important differences between white people and the Hopi. The Hopi believe that God is Nature. The Sun is the father of all people. The Earth is the mother. All Hopi life centers around the mother image. Clans and families are matriarchial. A child has three mothers: Mother Earth, a corn mother, and a human mother. The wombs of all three are identical. Mother Earth provides the religious shrines and ritual objects for our ceremonies: feathers, gourds, and cornmeal. Our religious shrines and places are not portable like the white man's God and heaven, that exist invisibly in the sky, and bless people instead of Nature. We believe the Earth is sanctified; where we live and farm is holy ground. It is the home of Gods and spirits. The white man places God in heaven and hell on Earth and it's been that way ever since he arrived. The white men sanctify themselves instead of Nature.

Hopi religious symbols are womb-like and of Nature. They are circular, protective symbols and concepts of life. The white man's religious symbol is the cross, which is shaped as a sword, or an erect penis. The white man's metaphysical symbols signify the rape of the Earth and its people. You live a crusade against Nature...

The Hopi ceremonial cycle is based upon the ongoing history of the Hopi people. It is alive and current because it is not written. The white man is entrenched in religious events of the Bible, written words that are outdated by thousands of years but are still taken as gospel. The Hopi practice and develop their religious beliefs all the time. All aspects of daily life are religious occurrences, not just a once-a-week lecture like those of the whites. Everybody participates in Hopi ceremonialism. It is not the word of one man. It is the spirit of the community.

Hopi heaven is life today. When we die our spirits come back to Earth and bring the power of Nature and of the Hopi to help our people maintain life. They dance in the plaza of our village. White man's churches are shaped like a phallus—spires and arches toward heaven. The Hopi kiva is in the Earth, circular and womb-like. Our ceremonies ask Nature to love us, to give us rain and crops. Every aspect of life—soil, rock, animals, water, weather, plants, *everything*—has a *living spirit*. We ask all of them to help us live. We respect and need the spirit in everything on Mother Earth. When our children are born, they are reverently kept in the dark for three weeks, then brought outside at predawn to gently greet their father the sun.

The Hopi do not proselytize or crusade. We have no insecure feelings that we will be attacked by those of other religions. We have no need to change others' religious views. Hopi religion would not fit others; others live in the wrong place. Hopi religious areas are in Hopiland. The white man sees the people of Nature as being the people of Satan who have to be saved.

Whites operate their missions under the power of the sword and of their machines. But the Hopi know that we live here and now. Our home is the Earth and we try to protect it and all living things. We pray to Nature's forces to cooperate with them and help them live. We go barefoot in the spring so that we might not hurt Mother Earth while she is pregnant. Our ceremonial dances are danced in soft moccasins so as not to harm the Earth. We live our religious beliefs. We love the Earth and we revere her; we think with our hearts.

Our white cultural perceptions have superimposed themselves on the blueness of many Native Americans. Because they have been to some degree assimilated into the mainstream of American thinking, these people can be as prejudiced against Nature as anyone else. We spoke with Passamaquoddy Indians in Maine who were planning to go back and live in the old Indian way in the woods, to hunt and fish and live off the land. They planned to do so armed with the technology of guns, fishing gear, and metal traps. They did not believe they could exist without these technologies, and admitted they had not stopped to consider whether the forest ecosystem could exist with them for any length of time. Establishing equilibrium was not their primary goal any more than it is the goal of most white people.

Indians have been asked why they first accepted white man's tools. Some have responded that these technologies have more power and magic over people and Nature than do the traditional Indian weapons. That is the seed of the myth of womb euphoria, and it suggests that prejudice against Nature is not isolated to the western culture. Along with the gun came the spirit of the lifestyle of which it was a part, a culture whose technologic euphoria captured many Indians with no fight at all.

Lame Deer:

> Buffalo are smart. They also have a sense of humor. Remember when we were together last in the Black Hills? When it suddenly snowed after a very hot day? Those six big black bulls we saw near Blue Bell, just like six large pick-up trucks. They were so happy over that snow. Gamboling, racing around, playing like kittens. And afterward we came across the tame cattle, hunched over, miserable, pitiful. "Moo, moo, moo— I'm cold." The real, natural animals don't mind the cold; they are happy with the kind of fur coat and galoshes the Great Spirit gave them. White hunters used to call the buffalo stupid because they were easy to shoot, weren't afraid of a gun. But the buffalo was not designed to cope with modern weapons. He was designed to deal with an Indian's arrows.

There are Indians today who have strenuously objected to the inclusion of the quotations from Lame Deer in this book. They feel his words promote a stereotype of Indian as a hunter-gatherer, a stereotype that does the *modern* Indian more harm than good.

Even Lame Deer himself had his inconsistencies. He seems proud that his father owned over two hundred horses. Yet horses are not indigenous to the plains; they are a *technology* originally introduced to the hunting-gathering Plains people in the late 17th century by the white man. And across the Rocky Mountains, several Great Basin Shoshone hunting-gathering and farming tribes also mounted horses and changed their lifestyle to include the large-scale hunting of buffalo. Within a short time, the Great Basin buffalo herds were exterminated, again illustrating that the euphoria of new technologies is able to override existing beliefs, cultures, and equilibrium with Nature. We have been told that the extermination by Indians of a wild buffalo herd has taken place in Canada within the past decade.

§

The Expedition has spoken with people who have related that the environment of many countries in the Eastern Hemisphere is in worse shape than ours. Many of us were surprised by these reports, because the religions of eastern countries preach oneness and equilibrium with Nature. We learned that there, as here, the euphoria and power of materialism have prevailed over the actualization of religious and philosophical teachings. In some of these countries, the result of prejudice against Nature is practically enslavement to a lifestyle of continuous work and recycling, in and of an environment without wildlife or natural areas. The admirable recycling technology they have developed there has stabilized the ecosystem just short of human starvation and annihilation of all Nature as we know it. (Perhaps it's nostalgia, but life without wildlife and natural areas leaves a great deal to be desired. And a scientist recently told us that countries with healthy natural areas use less energy to produce their gross national product. This suggests that Nature is doing a great deal of the work for us at no charge—which should make harmony attractive even to those who believe their heartbeats are sustained by economic gain alone.)

Our environmental problems are not caused by one single cultural belief. They emanate from a psychologically warped collective consciousness that is prejudiced against Nature.

Many believe that the word *prejudice* is useless in a responsible investigation because it is an emotionally charged term of daily language. To the contrary, that is precisely its strength. Too often, many scientists and educators forget that people are the personification of prejudice, that people are living symbols of centuries of prejudicial upbringing, outlooks, and ignorance that have greater influence and accessibility than the facts hidden in elitist terms and definitions. It is emotional recognition and correction of our habitual prejudices that are necessary to mend our environmental trespasses.

If permanent headway is to be achieved by our civilization, if as a culture we are to survive, we must deal with our prejudice against Nature.

How can we, as Americans, combat prejudice against Nature?

People only begin to think about problems when they are confronted by them. Confrontation can mean controversy. Although, at times, confrontation is disquieting, it can be a positive way to help people into awareness. The application of the Whole Life factor is a form of confrontation.

For example, a problem exists because some religious teachings seem to be prejudiced against Nature. Although for many people religion has been the philosphical scapegoat for our environmental dilemmas, religion is only caught in the same problem of prejudice against Nature as is every other of our institutions. Yet religion is close to Nature in many respects. *Nature is the art of God,* wrote Dante, and Genesis tells us that Nature was the first creation of God. But religion is powerful because it is complete. It is whole, in that it includes people's emotionality.

Any religious sect that desires America's young people to know God through the public school systems can get extremely close to that goal by insisting upon a curriculum that includes the Whole Life factor, the full study of Nature, and why people are prejudiced against Nature. Such studies close the gap between science, religion, and public education; they also assist the social sciences in limiting and directing runaway technologies that threaten to overwhelm our way of life. They fulfill the scriptures which say: *"Let us therefore cast off the works of darkness and let us put on the armour of light."* (ROMANS 13:12) (Note: Darkness and light both are equally valuable fluctuations of Nature; even the scriptures contain prejudice.)

I have designed what I call the Ten Amendments, and I tell people God gave them to me one day as I was skiing alone in the thin air of a high Colorado Rocky Mountain ridge. The amendments are a Whole Life factor confrontation of prejudice against Nature in religion. Those who can work at accepting them approach a religious harmony with Nature.

1. In the beginning there was God, who created the heavens and universe of her/himself, and called it Nature.

2. As God is alive, so is every aspect of the universe alive and equally blessed, be it animal, vegetable, or mineral.

3. God created people in her/his own image by giving people their own star (the sun) and solar system which was set aside to evolve and sustain people in equilibrium with all life and with God.

4. Each entity in the universe has been ordained to cooperatively exist with God and every other entity, for they are all one in the same, they are God.

5. It is God's will that all entities multiply and explore different ways of enjoying their God-given gift of *being.*

6. Each entity is given the freedom and responsibility to choose the manner in which it will cooperatively relate and survive in equilibrium with the universe.

7.   God sanctifies and blesses each entity's special gifts that it has evolved to sustain its own life. God's special gift to people is the ability to relate through consciousness and communication.

8.   The highest use of consciousness is to understand God and Nature and approach equilibrium with them.

9.   To gain full consciousness, relationships must be measured by their actual effects upon the solar system and all its inhabitants. It is a sin for people to not fully utilize their consciousness and instead act upon dogmatic statements allegedly emanating from God or any other entity.

10.   People have always had God's blessing to measure the millenia by different time standards. On some scales the year is 365 days long; in others it is 40 days in length; still others have years that consist of millions of centuries. All are correct in God's eyes. *A thousand years in thy sight are but as yesterday when it is past.* (PSALMS 90:4)

There are several aspects of these amendments that go against my upbringing and conditioning. They sit hard with me, but the years have brought me to the realization that there were many unholistic myths in my own upbringing, and that I have to relate to people and Nature, now, from a common trust point of view. *"Whatsoever a man soweth, that shall he also reap."* (GALATIONS 6:7) This seems to be applicable to our relationship with Nature.

Belief in the accuracy of my images and words was another of the myths with which I was reared. There are more to words than meets the conscious mind. Each word is a powerhouse, emanating from some ancient attempt to cope with the environment. Words not only serve to convey thoughts and feelings but when emotionally conveyed are meant to stimulate certain behaviors or mental processes in others. An inaccurate word stimulates incorrect symbolizations and concepts, elicits inappropriate feelings and responses, and leads to unfulfilling actions and relationships. The strength of language and cultural feelings led us out of the wilderness and into our present environmental chaos.

Symbols touch every part of our lives; they are tools of culture, and tools of education.

We must begin to symbolize that the Earth is alive, and that people are Nature. We must teach that we have two mothers, that the infant's relationship to the womb is a replica of an adult's ultimate relationship with the planet, that the newborn baby—quickly aware of Nature and experiencing the ecosystem within itself—should be encouraged to treasure its natural feelings, and that degradation of the environment is a form of self-hate. To attain social and global maturity, we must symbolically educate ourselves to cherish the feelings of life within ourselves, and cherish the planet which is that life (See Appendix B & F):

Lame Deer:

To our way of thinking the Indian's symbol is the circle, the hoop. Nature wants things to be round. The bodies of human beings and animals have no corners. With us the circle stands for the togetherness of people who sit with one another around the campfire, relatives and friends united in peace while the pipe passes from hand to hand. The camp in which every tipi had its place was also a ring. The tipi was a ring in which people sat in a circle and all the families in the village were in turn circles within a larger circle, part of the larger hoop which was the seven campfires of the Sioux, representing one nation. The nation was only a part of the universe, in itself circular and made of the earth, which is round, of the sun, which is round, of the stars, which are round. The moon, the horizon, the rainbow—circles within circles within circles, with no beginning and no end.

To us this is beautiful and fitting, symbol and reality at the same time, expressing the harmony of life and nature. Our circle is timeless, flowing; it is new life emerging from death—life winning out over death.

Hunter-gathering peoples, like animals and plants, experience life in Nature as a continuation of the prenatal womb. They have no other choice, for their cultural consciousness is lacking the technology to produce artificial wombs. They learn to know and respect the womb euphoria of Mother Earth and know the joy of sustaining an equal relationship with her. Adam and Eve were such people. The hunter-gatherers and rudimentary farming peoples are their most recent counterparts.

The mutations and environments which produced the strong imagery, symbols, and technological ability within American people triggered prejudice against Nature. Our symbols and technologies are not congruent with Nature. They are the apple given to Eve. They allow people the ability to carry within themselves another world, a fantasy world that artificially manufactures the euphoria of their childhood wombs. The early womb memories that are alive in each of us are the seed of prejudice against Nature. The ultimate delusion of humanity is the subconscious belief that the euphoria of the prenatal womb is attainable through techology and is more satisfactory than the eternal womb of the universe.

Womb euphoria delusions ignite the flame of prejudice against Nature, because Nature constantly interrupts the delusions with the realities of life. For this, Nature is resented. We counter by annihilating Nature with technology. Life becomes a matter of gaining technology instead of being alive. The joys of relating to the planet are lost in the fever to gain technology; the results are isolation from other people and life forms, competition, insecurity, anxiety, loss of identity, theft, pollution, chemical addictions, habitat destruction, species extinction, and war. We are the only members of the animal kingdom that regularly experience these problems. We are probably the only member of the animal kingdom that is prejudiced against Nature. No other life form seems to face the choices we face through the symbolism, technology, and religions of our culture.

Each negative symptom subsides as a person becomes emotionally conscious of his/her prejudice against Nature. Working toward equilibrium introduces the individual to values of life previously overshadowed by technology and competitive relationships based on power to acquire technologies. A new integrity evolves. It is reflected in one's personal outlook, and in an environment more able to support us. We must begin to work toward harmony. Nature would say, "You must adopt and teach harmonious meanings to your static symbols, to the words *Nature, Earth, standard of living, congruity, competition, technology, religion, and culture.* You must learn to apply the Golden Rule to the planet, for it is the only member of its species, and you have endangered it. You must learn to enjoy the sensations of my pulse and fabric. Try to increase your contact with trees, climate, rain, and clouds. Learn to love me for what I am, to appreciate the gift of life, which is the joy of us in concert."

Perhaps the most controversial and stressful confrontation between the forces of Culture and Nature takes place in the area of population control, an area which many people think is the key environmental issue that we presently face. I believe that my own experience with the problem deserves some attention.

When Diana and I married, we fully planned to have children of our own. But as we established the expedition school, we came face to face with the unhealthy environments and attitudes toward life that pervade America and its young people. We became less and less sure of our own desire to purposefully bring more children into a society that so brutalized life in order to achieve distance from Nature. We noted, for example, within the combined lifetimes of our parents and ourselves, a reduction of bird populations in the south Florida Everglades has taken place, as well as the extinction of all too many other forms of wildlife. We have personally witnessed during the past decades the rise of environmental laws and awareness accompanied by a further 30-50% reduction of Florida bird populations during this same period. We have witnessed an increasing mass-culture emphasis upon economics and standard of living, no matter its cost to the ability of our continent to support most forms of life including our own. Under these circumstances, we have come to feel it unwise and improper to bring our own children into this setting, and we, therefore, have refrained from doing so.

By bearing witness to the real cost of being Americans, our self-preservation feelings have been brought into full play. We also have noted that others who are aware and concerned about the unhealthy trade-off that our schools and economics demand, have limited their families, or only extended them by adopting children. I believe statistical studies would confirm that there is no population problem among those people who have let the plight of Nature reach their cultural consciousness. I believe that consciousness of the extinction, pollution, and social imbalance in our society stimulates alarmed self-preservation feelings which, in turn, instill population limitation *on an individual family basis.* Native Americans and other subcultures have told us that they practice natural forms of birth control when it appears to be appropriate. CNS can produce population controlling congruency that seems otherwise to be unobtainable on a mass scale. It is a consciousness that can be attained through awareness and concern about our own prejudice against Nature and the Whole Life factor.

Awareness of prejudice against Nature places a new fork at each step along the road of life. At first, one doesn't recognize the forks, but they are always there. They are the environment adjacent to each side of the road. Like any road, the road we presently travel is a cultural annihilation of Nature. Our roads are built to accommodate a wheeled technology that transports us to a fantasy civilization of womb euphoria. With awareness of our prejudice and our prejudiced conditioning, we can personally and collectively find new directions toward wholeness.

We can't all go on the Expedition, but we can begin to poke holes in our culture to allow for the voice of Nature to speak more clearly and strongly through us all. On the Institute, we've explored how this could happen, and we've come up with the ideas below. You can probably think of a lot more. We'd be happy to hear from you about your thoughts and experiences.

Measures that can be taken to prevent prejudice against Nature are similar to those used against any other kind of prejudice. By combining the methodologies of the environmental, mental health, women's, and civil rights movements, humanity may still have time to recondition itself to the reality of Nature's equilibrium. We can yet dance in the sun. An outline of basic measures to be taken against the discrimination of prejudice against Nature would include:

BASIC MOTIVATION: Self-preservation. Prejudice is outrageous. Get the rage out in the open. Let its energy work to decrease prejudice against Nature. Increase global thinking and local action by applying the Whole Life factor in all areas of endeavor.

WHOLE LIFE AVENUES OF CONFRONTATION AND CHANGE:

Personal:

Use your own creativity to determine directions and actions that holistically thwart prejudice against Nature as you discover its presence in your personal life. Consult with others who are also concerned.

Political:

Ask your legislators to introduce legislation calling for the use of a numerical Whole Life factor on all products and services sold in your state. (See Appendix A.)

Adopt an amendment to the Constitution, guaranteeing *all* species the right to grow and establish equilibrium with the planet. This would include a constitutional Bill of Rights for all living things.

Become active in local and national conservation organizations—place pressure on public officials.

Make Dr. Seuss' Lorax an anti-discrimination hero image and value symbol for Nature. Through him, confront and eradicate the inaccurate myths held by government, or other institutions.

Pass equal opportunity laws providing equal time in school for environmental education, contact with Nature, and encouragement of the expression of feelings.

Conduct mandatory on-the-job training, informing people of the environmental impact and implications of their jobs and professions.

Make populated areas legally responsible for the ecological soundness and maintenance of the natural areas needed to support them.

Make full use of city areas as habitats and niches for natural systems, gardens, and shrubs.

Have the E.P.A. verify each piece of legislation to certify that it is congruent to the way the natural world works.

Educational:

Teach awareness of prejudice against natural and global thinking as a central concept in school curriculla. (See Appendix B.)

Teach survival by reinforcing self-preservation feelings and consciousness.

Teach that planet feelings are important and worthwhile.

Design curriculla that explore and understand the Earth as a living organism in a holistic manner.

Revise subjects to include T-R (Tension Release), PAN (Prejudice Against Nature), CNS (Culture, Nature, Self-preservation), WL (Whole Life), and Distance-from-Nature factors.

Teach experientially, in outdoor settings. Don't teach *subjects,* teach self-preservation. Have written work used as letters to public officials.

Use the *Little Green Book* by John Lobell, as a required text.

Center curricula on the concept that people are Nature, and on an understanding of the inaccurate myths we harbor about Nature and how those myths affect us.

Have students design, for credit, programs that confront prejudice against Nature in their local areas.

Have students learn how and when to make their feelings objectively work to change the environment.

Have students analyze people, places and procedures with respect to their distance from Nature, and how to decrease it.

Identify how Nature in people is thwarted and attacked in the same way that it is in the ecosystem.

Use feelings attached to modern images in order to communicate Whole Life concepts. For Example; conceptualize *E.T.* to be an endangered species of natural creature; then show how *E.T.* gained protection and survival because people (kids) acted out their self-preservation feelings and concern for life and warded off incongruent cubic-red forces. Use Lassie, Flipper, Black Beauty, and others in the same way.

Make environmental education unmediated, provocative, and holistic.

Make environmental education experiential by placing students in situations where prejudice against Nature is harming animals, people, or the biosphere; give credit for these experiences as a student acts to reverse them.

Visit a mall to observe where you can find Nature, womb euphoria, self-preservation, or survival goods. How is distance from Nature being sold to the public there?

Organize student government committees to confront curriculum materials and teachers that are prejudiced against Nature, and that convey myths about Nature.

Make the last two years of high school optional and establish alternative, experiential Youth Conservation Corps-type anti-PAN programs for credit.

Develop classes where students and adults locate and trace the history of the suppression of their internal Nature and have them locate, identify, and attempt to correct similar situations in the ecosystem.

Recreational:

Fully develop knowledge and sensitivities for recreation that enhance feelings and Nature and discourage technology and competition; encourage wildlife observation, gardening, sanctuary maintenance, and non-competitive games.

Technological:

Devise technologies that are congruent with natural systems and help approach equilibrium; phase out unnecessary technologies by applying the WL factor. Investigate alternatives for each technology that is presently in use.

Medical:

Reduce birth trauma and technological birth rituals.

Cultural:

Encourage low-technology skills and arts which enhance Nature and celebrate life, such as singing, dancing, and non-competitive athletics.

Make the WL factor a numerical evaluation from 1-10 (or 0-100%), and apply it to all products and services, with 0 being environmental catastrophe and 10 being environmental congruency. (See Appendix A.)

Population:

Devise birth control devices that are medically safe and promote attitudes that enhance equilibrium on a global scale.

Occupational:

Develop career opportunites and training in conservation, environmental, and wildlife fields; encourage home and organic farming.

Confrontation:

Fan the flames of controversy; use emotions and non-violent antagonism to get facts into the open. Practice civil disobedience in any situation that is operating uncivilly against Nature.

At school, home, or work, make each situation and topic become closer to Nature. Teach your children, friends, and relations to apply the Whole Life factor and to ask the following questions of their teachers, representatives, parents, and clergy:

1. Does what you are teaching me, or asking me to do, help keep the life system of the planet intact, or is it instead hurting it by further subdividing it?
2. How is the long term effect of this subject or situation going to help the planet remain alive?
3. What is your training or expertise in knowing what is best for the planet's life?
4. Do you recognize that the planet is alive?
5. Is this situation or subject seeking immediate economic gains and/or distance from Nature, or is it instead trying to support a lasting global life system?
6. Could the money, time, and energy being put into this situation or subject better help support life by being spent elsewhere? How? Where?
7. What is the Whole Life factor of this topic or setting?
8. Where is the value of self-preservation feelings in this subject or situation?
9. Does this situation or topic tend to repeat our mistakes of the past, or instead does it help correct those mistakes?
10. Are Nature's and our life-preservation interests being fully represented in this topic, or is it instead being prejudiced against Nature?
11. If you or I do not represent Nature's own self-preservation in this setting, who will?

12. Are you afraid to discuss your real feelings concerning the subject matter?
13. Do you feel that I can disagree with what you are saying—even on an exam?
14. Are my feelings on the subject equally important to yours?
15. What value does the experience or subject matter have for you on an immediate basis? On a long term basis?
16. (When all else fails) Can you really teach me what I want to know, when what I want to know is how not to be like you?

Financial:

Give grants to people who need financial assistance to fight acts of PAN, no matter where they occur.

Do environmental impact audits on all citizens and have their taxes reflect any over-use of resources.

Refuse to pay taxes on government projects that are prejudiced against Nature.

Let taxpayers choose to which goverment bureaus their taxes will be paid.

Identify ten critical environmental problems, and provide large grants to help solve them—the grants to be distributed by committees composed of 11 to 14-year-old children who understand prejudice against Nature, and use of the Whole Life factor as a guideline.

Confront advertising that is manipulative or prejudiced against Nature.

Counseling:

Design counseling methods that are congruent with Nature, where the emotionally adverse acculturation of feelings is recognized, understood, and rectified; where personality is identified by utilizing T-R, WL, and CNS factors.

Create environments in which feelings can be expressed without fear.

Theory and practice should include analysis of relationships with Nature and the effects of those relationships. Discourage competition. Help people express and use anger in order to approach equilibrium.

Have people apply the CNS and WL factor to themselves by designing a self-exploratory inventory of attitudes, feelings, acts, assoications, career goals, interests—and then analyze this inventory in light of the WL factor.

Relate everyday values to natural feelings.

Media:

Have the theme of living congruently with the living planet appear in all media materials.

Continually collect and convey examples of PAN no matter where they are found.

Volunteer Work:

Organize or join national conservation associations and be active in their local chapters.

Confront local anti-Nature acts.

Oppression:

Use *Mother Nature* as a concept to identify with sexist exploitation. Enlist the help of people who have experienced discrimination personally. Help them to help themselves by helping Nature.

Religious:

Reinterpret ecologically unsound or anti-equilibrium religious teachings. Promote the Ten Amendments. Have religious groups confront anti-Nature dogma.

Consumerism and the Whole Life factor:

Because we are a consumer society, the most promising major act that would offset the impact of our prejudice against Nature is the actualization of the WL factor for shoppers. Its application to products and services is explained in the following article by Trudy Phillips, a graduate student and instructor at the Expedition Institute.

Full details of this proposal and its implications are found in Appendix A.

Whole Life Factor
or
What's This on the Package?

A large yellow school bus, sometimes referred to as a one-room school house, is parked in front of a shopping mall. The Expedition students are taking a few hours to shop for groceries at "Save-A-Buck."

"Okay, Billy, what do we want to buy for dinner?" Billy and Damon approach the end of the aisle where many different kinds of spaghetti and sauces fill the shelves. They begin picking up the packages and reading labels.

"This one has all sorts of BHA and BHT, monosodium glutamate...I don't want to eat that."

"Take a look at these three. They're all the same weight, and the prices differ by only a few cents."

Billy picks up one of the packages Damon is referring to. "Look at this one, though. They wrap the stuff up twice. That's a ridiculous way to package something—what a waste of natural resources."

"What's that on the right hand corner?" Damon points to a black circle with the words 'Whole Life Factor' written in white along the border, and a globe inside. In the middle of the globe is the number six.

"That's odd," says Billy, examining the other two packages, "these have the same symbol on them, but different numbers." Just then, a white-jacketed clerk, with Save-A-Buck dollar signs on his lapels, walks by carrying his inked price stamper.

"Excuse me," Damon pleads, pointing to the circle, "we saw this symbol on these packages of spaghetti, and we were wondering what it means."

"Well, it has to do with the environmental impact of a product, how much that spaghetti costs the environment. It helps the consumer make a more environmentally sound decision when buying a product. Over there in that rack, there's a magazine which will tell you more about it."

When they reach the magazine rack, Billy and Damon both find the one the clerk was talking about. Each of them picks up a copy and begins to browse. As Damon leafs further into it, a smile forms on his face. "What a great idea! Finally, someone has put together a different kind of 'Good Housekeeping Seal of Approval,' one that deals with *Earth Housekeeping*.

"Here's an article on the different kinds of spaghetti we looked at. It talks about how they come up with the different numbers on those packages and the cost to the natural world in terms of transportation involved in marketing the product, packaging, whether or not the ingredients were obtained locally, and more.

"That means shopping would be faster and easier. Every time we were faced with the decision of which brand to buy, all we would have to do is to look at the Whole Life Factor. If one brand has the lowest impact on the natural world, I don't know why we would buy any other."

§

The preceding dialogue could be the outcome of an idea discussed during one of the Lesley/Audubon graduate courses this summer, *Approaches to Research in Environmental Education,* led by Dr. Jim Swan. This idea would educate the consumer as to what the products and services they buy cost the *planet.* It is easy to buy something with the balance of the decision resting on the impact on your pocketbook. Less often do we actually consider the impact on the natural world, the hidden costs, which in time translate into costs to the pocketbook. It is very difficult for most of us to make connections between our purchases and acid rain or Love Canal.

Although many people are aware of environmental problems, they find that translating this awareness and concern into consumer action is often impossible. An overwhelming array of facts and figures, viewpoints and technical jargon tend to create feelings of confusion and anxiety when people find themselves mired in this complexity of data. Eventually, some experience paralysis, which brings on a sense of decreased concern and commitment to their environmental awareness.

It was postulated that the Whole Life factor would resolve this by utilizing a numerical system. Numerical values are often used as a simpler way to express complicated concepts, such as the acidity of water, or the concentration of one substance in another. Thus, if environmentally concerned people were shopping, all they would have to consider would be the Whole Life factor. This new concept would take into consideration the amount of energy consumption, resource use, toxicity, packaging, recyclability, and likely environmental consequences of the usage of the product.

At the same time, a magazine could be published which would contain explanations of how the numbers were assigned, comparisons of products, and other articles and editorials related to environmental impact.

Presently a proposal is being written to bring together experts in various disciplines to discuss criteria for evaluating products and services. It is an idea in the beginning stages, a very exciting idea in terms of possible educational impact. Perhaps some day, like Billy and Damon, you too will be able to walk into a store and choose, from among different brands, the product which is more congruent to the way you would like to interact with the natural world.

If you have any comments or suggestions, I would appreciate receiving them.

§

It has taken our culture some four thousand years to produce the social and environmental mess we live in today. It may take us many generations to experience a living equilibrium with Nature, for we have burned a lot of bridges behind us. It makes no difference how powerful are our religions; how much money, status, and influence we have; how educated we can become; how scientifically and technologically adept we are; we never can bring back the Auk, the Passenger Pigeon, the Woods Bison, or the Heath Hen. They, with many others, are extinct. They have truly died. They were Nature's children; we exterminated them. The relationships, environments, and moments in time that produced them and were part of them, are long gone. At the present time, we are annihilating one species every day. Are we approaching the situation to which the scriptures refer? *Thy wrath is come...and shouldest destroy them that destroy the earth.* (REVELATIONS 11:18) (See page 259.)

But even injured, the Earth lives on, and will live indefinitely. The question facing us is not a matter of the length of the Earth's life; it is a matter of the possibility of our life. It is a matter of ethics, morality, and reason. Should we not apply our knowledge and our mental powers toward correcting our prejudice against Nature, rather than toward technologies that reinforce it? Of what use is the ultimate in technology, if it is not healthy for us? Our toes are being stepped on, and we are being taught that we are wrong to complain. Out culture is dominating our nature and self-preservation. Again some sound experience from the past: *If any man have an ear, let him hear.* (REVELATIONS 13:19)

There is something missing in the quality of life when, in the name of peace, we have stockpiled atomic bombs in quantity sufficient to annihilate life on Earth twelve times over; when moral leaders practice immorality; when environmental leaders crave high wages and technologies; when our civilization thrives while demeaning individuals, Nature, and other cultures; when competition, violence, and exploitation of people's insecurities are the basis of daily life; when the addictive euphoria of drugs, alcohol, and tobacco is a rampant sedative against feelings of unfulfillment.

I never have trusted people who have only one answer for everything. I trust the concept of prejudice against Nature as a single answer only because it raises an enormous number of questions about most of the other *answers* I ever have heard about life.

I am familiar with the kind of world that the established answers have fostered. I am not satisfied with it. It is healthier to trust in the unpredictability of Nature. Her wholeness is my home, and I am living well. I am now aware that further dependence upon our historic answers and approaches to survival and civilization can only intensify our global problems.

I mistrust the answers I have received from our institutions. I equally mistrust the questions that my culture has taught me to ask. Exploitive, anti-Nature questions can produce answers that are correct in the context of the culture, but are terribly wrong for survival of human life. We have earned good grades for learning how to hurt ourselves, and the situation continues to deteriorate.

In order for people to use their democratic privileges and take into their own hands the direction of their lives and communities, they must know where their strengths lie, and how to use them. We have found that using an awareness of our self-preservation feelings to counteract inherited prejudice can be, and is, a vital life goal. Individuals who try to offset their prejudice against Nature can establish deep, rewarding relationships with others of the same persuasion and with the Earth. All ways of life are worthwhile as long as they are striving to approach equilibrium with the main source of life.

We never again will be the hunter-gatherers that we once were; there are too many of us and the environment that made it possible once is now extinct. A reasonable goal is to be hunters and gatherers of harmony with Nature today. We still can reap the euphoric benefits of living whole lives. Money, technology, competition, and chemicals are not the only ways to obtain euphoria. Americans have a constitutional right to life, liberty, and the pursuit of happiness. They all exist in harmonious relationships with Nature.

Living in harmony with the natural world and with people is a different approach to life than the approach broadcast on television. It is anything but canned, anything but sterile, or distant from Nature's life-giving T-R fluctuations. It is a self-regulated expedition, an adventure, that makes one feel alive. Life is worth living because one has constant contact with the values of the life process and its intriguing fluctuations.

Approaching equilibrium is not a difficult path to take. As one learns to more deeply symbolize that the Earth is one's mother, that she is alive and nurturing us, and that all life has a right to live in equilibrium, one can find and actualize energies and directions that never seemed apparent before.

To begin to live closer to Nature and to love her because she is your present mother and womb-replacement immediately initiates the seed of euphoria and personal equilibrium. If Nature is one's present mother, and if we are discriminating against her—starving her, poisoning her, tearing her apart piece by piece, grinding her up for a profit or a hobby—then it would seem fit and right to try to do something about it on every level that works, if for no other reason than that we're hurting ourselves in the process.

At the end of a discourse such as this, the appearance of a synopsis or conclusion is always helpful. Because of the circumstances surrounding the creation of this book, its conclusion is actually found in the prologue and in Chapter 4. It is an interesting circular, and appropriate situation, because in this manner its cyclic quality emulates Nature's ways. I strongly encourage the reader to go back and reread these sections. Will they read differently now that their full stories have been told? Sometimes I think that when walking in the sun, some of its light enters one's mind. Each of us is entitled to use that natural brilliance to examine our heritage: our past and present relationships with ourselves, our civilization, and Nature. Inside each one of us is contained a weaving solar-powered searchlight of consciousness and reason that randomly illuminates within us what we are, and what we have been made to be. As its light bounces off our cultural heritage, we are able to discern the cubic-redness of the Western world. But when the searchlight's golden sunbeam strikes our blueness, we experience green. The golden beam blends with our blueness because it is of the same origin, it is Nature.

As Americans we have the freedom to choose either to acknowledge or to ignore the gratifying wholesome green blend of reason and Nature that exists in our awareness. But if we are concerned about our self-preservation, it is not simply a matter of choice. It is a matter of survival, and a matter of time. It is a responsibility of life to create gratifying personal and cultural relationships with Nature that are green, that are in tune with the hills, the wind, and the stars.

A picture is worth a thousand words.
An experience is worth a million.
An action *is*.
Let us subdue our prejudice against Nature.

§

Afterword

Sharing The Good News

by Jim Swan, Ph.D.

co-author of SWAN and STAPP: *Environmental Education:
Strategies Toward a More Liveable Future.*

On the evening of March 21, 1970, I sat in a small room upstairs in the Union Building at the University of Michigan with three other people, one of whom was the noted ecologist, Dr. Barry Commoner. Just a few hours earlier, 15,000 people had filled Chrysler Arena there to hear speeches by Commoner, Arthur Godfrey, Senator Gaylord Nelson, and representatives from half a dozen student groups, and music by Gordon Lightfoot. By a scheduling quirk, The University of Michigan Earth Day 1970 Teach-In on the environment was held one month prior to the April 22 date observed nationwide. That made this the first large-scale environmental teach-in in the United States, and Commoner was elated. It was a sweet moment of victory for people such as Barry Commoner, Margaret Mead, Kenneth Boulding and others who had long been environmental advocates. The standing ovation for David Brower that night signalled that finally the nation was willing to listen to the pleas of the ecologists who kept telling us that the earth was in trouble.

We who had organized the teach-in were talking excitedly about the evening and the events of the days to come, when a moment of silence came over the small group. Seemingly struck by a profound insight, Commoner turned to me and said, "Jim, I don't think any of us here tonight are fully aware of the implications of this teach-in, or what it may ultimately mean for society and the world." We all sat there in silence on these words. This was the first of the large environmental teach-ins of that first Earth Day in 1970, which brought the issue of environmental quality into full public view for perhaps the first time in modern society. Yet it was as Commoner observed, a little like a person in psychoanalysis having a profound insight and then having to work through the implications of the insight to really grasp its meaning.

It was an especially important experience for me, though at the time I didn't realize it. As a young faculty member at the University of Michigan, I had recently sat through a long series of discussions led by Dr. Bill Stapp,

233

which led to the creation of a working definition of environmental education that has been used since then by countries around the world to help them formulate their environmental education policy. The goal statement to which we had agreed in that seminar was that environmental education was concerned with developing a citizenry which is aware of the biophysical environment and its associated problems, understands these problems and how to become involved in resolving them, and is motivated to do so.[1] In contrast to the earlier movements of outdoor education, nature study education, and conservation education, with their strong emphasis upon nature appreciation, recreation, and natural history, this new definition took the bold new direction of specifically addressing environmental *problems* and finding ways to get people involved in resolving them.

The gnawing question which arose from these discussions and this new definition was what to teach in the name of environmental education, and perhaps more importantly, how to teach it. This, I think was the theme that Commoner was catching, which added a note of sobriety in the Earth Day euphoria. His insight was later supported by some research done by Dr. David Lingwood, then of the Institute for Social Research at the University of Michigan.[2] As a professional people-watcher, Dave and his research assistants were out with their questionnaires to poll the crowd at this first Earth Day Teach-In. He found that most of the people there were already knowledgeable about environmental problems. They had come to *vote with their feet* for attention to pollution, resource depletion, and associated problems. What they wanted were solutions.

This said a lot for the educational efforts which already had been done; but what about the people who *weren't* there? Earlier, after noting that at least initially the environmental movement was a white middle-class phenomenon,[3] I had done studies of environmental awareness and concern in the black community of Detroit's inner city. My research showed that some of the students most well-informed about air pollution did the least about it, because they felt it had a low priority for them in contrast to racial and economic issues. This research and Lingwood's together strongly suggested that simply informing the public about problems might not be enough to produce much change unless other factors also were considered. This is not to say that informational programs are bad, but simply that you tend to reach very few people who aren't already predisposed toward an issue, unless something unusual is afoot.

Since Earth Day 1970, I have been on a personal quest to find answers to the issues raised by the first Earth Day, and Barry Commoner's words have often come back to me. First I ventured into an investigation of values clarification. This process gradually led me to train to become a humanistic and transpersonal psychotherapist.

For a time, I became so involved with psychology that I had just about given up on environmental education. In reading about the healing work of American Indians, my interest in person-environmental relations was rekindled. In earlier times, they sought to heal by increasing harmony with nature. Fortunately, I was able to study with two medicine men, Sun Bear and Rolling Thunder, and to witness firsthand the sometimes miraculous spiritual healing they do. How can they apparently heal a person by waving an eagle feather or creating a circle of stones and praying and chanting over them? When I asked, instead of describing techniques, they would talk about learning to live in harmony with nature. They would quote people like chief Joseph of the Nez Perce who said, "The earth and I are of one mind," and talk about learning to "walk in balance on the earth mother." The meaning of these words for mind, body, and spirit is still a process which remains largely a mystery to me. But hearing them, I was reminded that Nature is perhaps one of the most critical issues of our day and that environmental education is a key to our society's understanding of its true relationship with Nature. This is why I wanted to visit the Expedition Institute.

Both Mike and his wife Diana stressed that I should familiarize myself with the Audubon Expedition Institute program *before* I met with the school where I was to teach a class on education research methods. This I began to realize as I read copies of *Our Classroom Is Wild America* and *Across the Running Tide*. AEI is an experiential program designed to critically examine and deprogram much of the rote learning that goes on in other schools and to help people to become autonomous. It is an accredited high school, college, and graduate program geared to help people holistically and academically to get in close touch with nature within and without, and to explore ecologically sound lifestyles. This I liked a lot. They were not only talking about things that I believed in; they were doing them. It seemed like a fresh breath of air, and so I was off to the woods and waters of Maine.

Bouncing down a gravel road in rural Maine, Mike began to orient me to the class ahead. "They've been reading the book you and Stapp wrote," Mike said, "and they think that it needs to be updated." I agreed that the book was six years old and probably could use some reworking, but little did I realize what I was about to walk into. The students in the Expedition Institute and I met the next morning at the classroom building named the *Greenhouse* (because it is painted green). Mike had invited me to be informal, as that was the atmosphere of the group. These people had been literally *on the road* for two years, travelling around the country in yellow school buses, sleeping out in tents in all kinds of weather, and exploring first-hand what living in harmony with nature means to different people in different parts of the country. They had been to the Hopi ceremonies in the Southwest, lived among the Amish of Pennsylvania, and studied the folk customs of New England. They had travelled from New Brunswick to Florida and westward from the Hopi villages to the Olympic Peninsula. First-hand, they had studied a combination of archaeology, anthropology, sociology and ecology as a unified experience all the while living as a close-knit community. Also generated in this living-learning community was the impetus to develop an ecologically sound lifestyle. This is created out of group discussion and personal experience. Everyone seems to become responsible to everyone else; at least this is what I saw.

I found that my first encounter with the group was just that—an encounter. They knew me by a volume I had written six years prior. I knew them by the books Mike had sent me and through his briefing in the car between the Bangor airport and the land near Lubec where they have their base camp. I had not expected the students I met. Some of them were older than the average graduate student. They had some experience in the world. Several had been school teachers or naturalists. As I walked up to the Greenhouse that first day, I heard a string band playing. During the course of the year many of them learn to play an instrument: a guitar, mandolin, banjo, or at least a penny whistle, as well as to sing the ballads of the land. A stop at Pete Seeger's sloop, Clearwater, on the Hudson River is an annual event. They really do know something about harmony, I thought to myself, as I approached the Greenhouse that first day.

The charge of the class to Lesley College, the accrediting institution, was to present the fundamentals of research methods, enabling students to formulate research projects, as well as prepare them for additional classes on research methods, if they wanted to go on for a Ph.D. After the first few *get to know you* exchanges came the first question: "What do you think about the existing environmental education programs in this country?" The ensuing discussion on this one question became our first class of three hours. By the time the period had formally ended, we had sheets of paper on the walls listing as many different programs as we could think of. I was not teaching a class; rather, I was more of a rudder, steering a boat that was already underway by asking it which way it wanted to go. This atmosphere remained with us for the next ten days. There was never an issue about class participation. It was almost as if the energy was there waiting to be voiced; I simply facilitated the group as an exploring unit. The experience of being on the road for a year or two had primed people for the discussion that took place. We became fellow members of a ship looking for a new land, or at least for a new way to get to the existing one. The destination, they said, was the kind of relationship with nature that people like the Hopi Indians or hunter-gatherers have. I couldn't argue with that, but then challenged them to state that as a goal. What we then came up with was a definition of environmental education which really is radical, at least in contrast to what I had written down a few years prior with Bill Stapp. *Environmental education is the process which aids humankind in developing behavior congruent with maintaining the earth as a living organism.* This is the goal we came up with. I stood there in front of the class looking at what I had written down. Then I turned and almost paraphrased Barry Commoner: "Do you realize what you're saying?" I asked. They responded that to state anything less was not living up to what was needed. I couldn't argue with them. In my work with the medicine man, Rolling Thunder, I've often heard him declare that, "The earth is a living being, and we should treat it that way. You learn to love and respect the earth as you would ask another human being to treat you. This is how the Indian people have always been brought up to think."

In my years of working with Rolling Thunder and Sun Bear, I've come to accept this as true. It feels right and comes out right, and what else is there when it comes down to discovering what is true in the world? Yet it's the sort of thing that said publicly has gotten me in trouble, because it is a view that is foreign to our traditions and to those who reinforce them.

I used to be a college professor of environmental studies. Then I began to explore human potentiality and American Indian ways. In the process, I changed and I tried to relate this to my students and colleagues. The students liked what Indians and humanistic psychologists like Abraham Maslow had to say about living in harmony with Nature, but many of my colleagues were threatened by it. Maslow found that self-actualized people tend to have a deep reverence for Nature. Yet many of my academic friends felt this was *wishy-washy thinking.* When Mike told me that the title of this book he was writing was *Prejudice Against Nature,* I was really hooked. I couldn't agree more with his analysis. It seems like so much of what humanistic psychotherapy, meditation, and American Indian ceremony and ritual does is try to poke holes in our prejudices against nature, so that the culturally blocked connections between nature within and nature without will become unplugged and resonate together in harmony. It's interesting that on one hand we desire the environmental benefits of harmony with nature; but on the other hand we get concerned about the processes of achieving harmony—if they differ from our cultural conditioning.

The problem with modern culture is just what Fritz Perls called *projection*—we tend to do to ourselves as we do to the world around us. The modern epidemic of psychosomatic illnesses that beset Americans, cancer, heart trouble, ulcers, etc., all are due to denial of our inner voices— our own inner nature (Nature within). While subduing nature around us, we poison ourselves with a myriad of chemicals in our foods, and the air we breathe as we dump these chemicals into the environment. We must be desperate to continually try to create ways to control and alter the balance of nature to do this to ourselves. We synthesize a new chemical compound every minute. Rather than seeking to sensitize our minds and bodies to nature and learning to live in harmony with nature's subtle pulses and flows, we seek to control nature. We belch clouds of wastes into the air in the process of walling ourselves off from nature in concrete structures which may do us more harm than good in the long run.

There is a profound neurosis in our culture and it is the fear of the future. Fear breeds prejudice—*prejudice against nature*.

The title of the book alone got my interest. Rather than trying to plan and regroup myself for the next day of class, I began reading the manuscript.

The setting for the first chapter is the tidal marsh, the predominant feature of the Lubec land owned by Mike and Diana. I went down to the rocks exposed by the receding tides (which are 27 feet in this arm of the Bay of Fundy) and began to read. I had just flown in from Seattle which is on Puget Sound, also a large bay with tides and salt water; so I felt quite at home. But in just a few hours I would marvel to see the uniqueness of this place. At one time, I'd look at this bay with three tiny islands in it; come back six hours later and it would be empty, exposing rocks and tidal flats that extend to the islands over a mile away. The surge of water through the estuary is like the life blood of Earth pulsating through an artery. The metaphor of the earth being alive took on more meaning.

Watching these tides, I decided to use that energy in class rather than to plan a lecture. Good research depends on careful planning at the conceptual stage. Scientific research is never objective. Because the unconscious asserts its influence, scientific methods *attempt* to be objective, but the choice of what to study is always a subjective one. Therefore, rather than fight the flow, the next morning I challenged the group to look at the assumptions underlying the conceptual definition of the earth being a living organism and the need to help human behavior become congruent with this working principle. To actualize and fully accept this definition, the group came up with the following statement of assumptions:

1. The planet earth is seen in its entirety as a living organism.
2. Nature and human nature are one and the same.
3. People are an organ of the organization of the earth.
4. All elements of the earth-organism system are interrelated.

5. Our culture, language, and symbols are presently inadequate to explain the earth as a living organism and that people are a part of that organism. Therefore, it is imperative that we develop functional communication which incorporates people, nature, culture, and the planet as interchangeable terms.

6. Sensing the totality of the earth as a living system involves sensing the wholeness of yourself.

7. The life of the planet is in continual fluctuation between tension and tension-release of energy; e.g., weather patterns, tides, thirst, and emotions.

Aside from the logical deductive process of justifying a new definitional statement, the items which seem most significant to me are numbers 5, 6, and 7. Whenever I have critically ventured outside the bounds of academic psychology, I have been confronted with the limitations of many conventional theories. Carl Jung, who is only now beginning to receive some recognition in academic circles, proposed that there are four elements of the human psyche—thinking, feeling, sensate-practical, and intuition. We are a thinking culture, almost totally embracing Descarte's notion of *I think, therefore I am.* The pathology of total focusing on thinking alone is to deny the rest of the self and become mechanical, and in the process, alienated from the rest of life.

Most of the new humanistic psychologies are concerned with freeing up emotions and intuitions which have been blocked off by modern society and education. Graduate school for most people is concerned with developing rational, analytical, deductive thinking. In contrast, Sun Bear, a Chippewa medicine man, was tutored by his uncles as part of his training to become a medicine man. What did he do? He did chores like chopping wood and carrying water from the well. Then at the right time he was supposed to stop what he was doing and go to where his uncles were waiting for him. If he was at the right place at the right time, he was given some instruction about the healing properties of an herb or some other element of his medicine training. There were no formal classes such as we offer. No reading assignments were required. He did the basic chores of life and had to learn to intuit to be at the right place at the right time—over a period of time, he learned the skill...or art. Later, they taught him to track game with his sense of smell. This education seems to be an equivalent to a graduate degree in intuitive and sensate-practical functioning. Think of how different this is from what we are conventionally taught. In most schools, we are taught the *three-R's* as basic skills—but what about all of our other life factors? Everyone dreams, but schools seldom help us understand our dreams. Everyone relates to other people in various kinds of relationships like expressing and understanding emotions, making agreements, and being understood. I couldn't agree more with Mike that nature speaks to us through feelings and with sensations and that our modern culture has a problem about nature, within and without. Emotions, feelings, intuitions, sensations (including your big toe throbbing right before a storm comes) all are part of still owning our personal connection with nature. Recently, we've been able to substantiate such anomalies as your Aunt Bessie's weather-sensitive toe as her sensitivity to positive ions in the air and changes in the barometric pressure as a front approaches. Sun Bear says, "People who live close to nature know when the earth changes are coming." This is precisely the purpose of the kind of education that Sun Bear and others have gone through. While some people are sitting in schools learning abstract theories, others are learning to sense and feel nature so they can sense the tides, the frontal weather patterns, and the responsive movements of fish and game. In the old days, such sensitivity to nature's moods sometimes meant the difference between life and death. They are aspects of self-preservation in nature. Mike talks about alcoholism and prejudice against nature. If you were an Indian, raised to value sensing and feeling nature and then thrust into a culture that says that such an education is worthless or even a negative influence on success, what choice would you have but to become depressed, feel worthless, and want to try to deaden the subtle voices you feel? The drunken Indian in the street may almost be a symbolic expression of our culture's prejudice against nature.

The assumption that our culture lacks the words and symbols to adequately communicate about the person-planet relationship is probably accurate and a good indication of what we have chosen to deny in order to develop our technological world. Simple, subtle things like words and expressions can tell you a great deal about a person or a culture. Mike's statement, "Our environmental problems remain and multiply because we are not conscious of our roots and symbols," hits home. In the American Indian culture, it was customary among many tribes for young men and women to go on a *vision quest,* where they would discover things about their true nature and their connection with nature. Usually this meant years of work in preparation culminating in an experience of going off by oneself, often to a place on a mountain top or in a cave, and spending several days in isolation praying and fasting. Visions might come while awake or asleep. Animals might come to the person. Feelings and impressions would manifest, like Mike's conversation with Mother Nature. Then the person would return to the community and tell the medicine man or woman what had happened. Then the medicine man or woman would give guidance on the person's life direction and work. Here, personal symbols and feelings were sought out as truth, for they represented direct statements from a higher power which seemed to blend people and nature together.

Increasing numbers of people have been experimenting with various conscious-broadening methods, such as drugs, gestalt, bioenergetics, meditation, yoga, and tai chi. We are seeing a new phychology beginning to emerge. In dreams, visions, and unusual experiences, we are seeing people popping into realms of consciousness for which they may have no maps. In the eyes of conventional society, psychiatry, and psychology, many of these people seem to be *out of control* in their experiences and are considered psychotic or schizophrenic. Because there is no scientific control or measuring device for their experiences, they are called *mystical*. If one understands the nature of mystical experiences, especially those which seem to unite people and nature, then what many experimenters are experiencing is the equivalent of the vision quest experiences which some cultures purposefully plan. Relating to these experiences this way, as positive experiences of a transpersonal nature, can often result in quick integration towards a very positive and more alive mental state. It seems that through our authoritarian years of pragmatic conditioning, we build up resistances to natural mystical experiences, and then for some reason we break through. The project designed to handle this phenomenon is called the Spiritual Emergency Network and is rapidly becoming an international phenomenon to allow for creation of a new consciousness that will let mother nature speak through us.*

It isn't necessary, however, for everyone to have a vision to become connected to nature. My own roots go back to the Eskimo people of northern Europe. They have a concept of God called *Sila*, which can be translated as the *Great Force Beyond the Seen World*. They say that Sila is not always present, obviously, but comes to us in time of need in our dreams, our failures, the laughter of young children, and through natural phenomenon. Mike and Diana Cohen's efforts are directed toward sensitizing people to the subtle voices of nature which the Eskimo people call Sila or at least the manifestation of Sila. In many laboratories, through the study of brainwave patterns, biorhythms, and the subtle electromagnetic fields of life, we are learning that we are not dense bodies, but rather that we are fields of atoms and molecules stuck together for a time. Because we are fields, we also are affected by fields around us, which explains Aunt Bessie's weather-sensitive toe. The important thing that seems to be different between ourselves and the Eskimos is how we feel

---

*The Spiritual Emergency Network is based at the Esalen Institute in Big Sur, California.

about the origin of this force. The Eskimos say that it is the pulse of God speaking through us, and they worship this life force. We call it weather-sensitivity and give Aunt Bessie some cortisone for her sore joints.

Some people seem to fear nature, because they can't control it. Because of their fear, they seem to get stuck in a couple of places in their acceptance of nature within. Either they try consciously to deny experiencing it, or they try to dismiss the importance of the experience. In either case, the denial of nature within has an external counterpart; it is our alienation from our natural world. If we then break down the barriers to experiencing nature and find that the inner and outer language is new, it is not nature's fault. It's our challenge to come up with a new language to describe this perception. Describing our relationship with nature as being prejudicial is a way of meeting the challenge.

Returning to our classroom at the Greenhouse, it became apparent that the group wanted to be a voice for a new definition of environmental education. Because their statements and arguments hit so many resonant chords within me, we continued from making this statement to looking at its implications in terms of validation. Having acknowledged and recorded the initial statement as well as the assumptions underlying it, I then challenged back: "Okay, we've had the guts to make this scientifically and academically outlandish statement; how do you translate it into measurable elements like skills, knowledge, and behavior? If you're going to communicate this philosophy to a nature center director or a school superintendent, you've got to get more concrete."

After discussions about methods of evaluation and defining terms common to educational research, they came back with the following:

> The translation of this working definition of environmental education into action involves attention to: skills, values, perceptions, feelings, behavior, and knowledge blended into a working wisdom of thought and action which is congruent with the definition. The components of this working definition are:
>
> 1. *SKILLS:* to be able to create and sustain consensus-based communities dedicated to personal self-reliance, continual growth, and self-actualization which arise from an increased awareness of the earth as a living organism.
> 2. *VALUES:* to construct an earth ethic which values personal responsibility to sense and respond empathetically to the living planet.
> 3. *PERCEPTIONS:* to develop a personal perception about the nature of the world, its connectedness within us,

and the resulting sense of the unity of all things.

4. *FEELINGS:* to generate an understanding of the planet's fluctuating tension and release as experienced in people and how their feelings are expressed culturally, thus encouraging the acknowledgement, expression, and acceptance of feelings which are congruent with the life of the planet.

5. *KNOWLEDGE:* learning how to learn a wholistic perspective of the life processes by encountering nature's questions and using them to lead the individual towards a working knowledge of life.

6. *BEHAVIOR:* to integrate attitudes, feelings, perceptions, knowledge and skills into an active lifestyle of maintaining the earth as a living organism.

Rather than going over this statement point by point, it seems more appropriate and vital to look at what it's saying and to note the process which led to its creation. Here are a group of people who did not know each other before they came to the Institute. They have gone through a series of global learning experiences which have changed them and ultimately have given them an incredibly strong sense of community. This definition was developed by the consensus of the entire group. To get any group to agree on such a radical thing is a near miracle in itself. Also of importance is the fact that they came to this group decision in a relatively short time, considering that they came from backgrounds as diverse as rural towns and the inner cores of large cities. During the time on the Expedition Institute, a true supportive community was formed. This enabled people to grow, change, explore and creatively exercise an integrated life process. Since a good deal of time is spent in group process, ideas about people, nature, culture, and how this relates to lifestyle had been hashed around almost daily during group meetings. They were not graded on reaching consensus. They found its value by experiencing it, by reaching it. That they did so easily is what to me seems important. Not only does what they are saying ring true to my inner ears, but the electricity of the decision-making process remains a fond memory. We spent a good deal more time on this class than was planned, simply because everyone including myself wanted to. This is when education is exciting! We were taking the thoughts and experiences of two years of work and giving shape to them as guidelines to explain the world and how to deal with it.

The people on the Expedition Institute program really have taken Barry Commoner's thoughts to heart. They are saying that things are not working well the way our civilization is set up. We have made some important gains on cleaning up the environmental problems, but there is still a long way to go, and maybe we started out by tackling the least difficult problems. You can pass legislation to control air pollution, but then along comes another administration with a different perspective and that same legislation is repealed. The laws and recycling centers help—but just what are we doing to improve the environment by changing around the consciousness of the culture to remove our prejudice against nature? As Mike has pointed out, the roots go much deeper.

Funding for environmental education programs is drying up these days. Is it because the programs are not valuable? I doubt it. It seems as though at the very core of our prejudice against nature is a fear of nature, and the experiences of nature within and without, which Mike has described so well. The description of shooting the woodchuck brings this out with a confrontation of death and killing. Animals kill to live, whether they eat plants or animals. In nature, killing is done out of need and not in excess. Man is one of the few animals which can kill with ease beyond satisfying needs. Lions and bears must hunt. A hunter with a rifle can kill off many animals without acknowledging what has been done. It is just like going shopping in a supermarket and forgetting that the packaged meat was once on the hoof. By ignoring death we ignore life. We insulate ourselves from killing and death and lose touch with the mystery and wonder of life.

I am not a vegetarian and don't believe that it is natural for most people who have evolved in northern climates to be vegetarians. But I do believe that we need to restore a sense of reverence for life, which includes an acceptance and recognition of life and death and the associated feelings. In deep therapy, continually we find at the very core of each person a sense of love and an embracing of life. If people rediscover this connection to the nature within themselves, then they can grow and develop toward self-actualization with a sense of wonder and awe about life that is beyond judgment and evaluation. It is simply a surrender to life itself and the experience of being alive. Descarte had it backwards; he should have said, "I am, therefore I think."

It seems to me that the students I met in the Institute program all could relate to what I have just talked about. Some of them openly talked about having been hooked on drugs or alcohol before joining up. Others talked about depression and feeling worthless. Somehow during the course of the Institute, they had broken through to their inner selves, which had been nourished by mother nature and a supportive community. This experience led them to come up with the environmental education definition upon which they all agreed. And they had done this without psychotherapy, taking on a religious devotional practice, or becoming freaks. Carl Rogers has always maintained that strong positive interpersonal relationships are therapeutic in themselves. This is part of what I think happens during the time spent with the Institute. It is something that the students seem to take with them.

Another thing which happens is that people seem to be able to learn to use the environment to educate themselves, as Mike talks about. I'm not sure how this happens; but on short field trips with some of the people, I found myself relating to places slightly differently as they guided me about rural Maine. Equilibrium, I think, is a key word for describing what the Institute teaches. They learn it academically, but also through contact with nature, sleeping out in tents in all sorts of weather and hiking through wild places. They also learn it through song and dancing folk dances. To really understand what the Institute is trying to do, you should attend one of the group sings or join them when they sponsor a contra dance in a rural old time meeting place. Get a copy of the *Equilibrium* album which they produced through Folkways Records and the National Audubon Society. You can learn about harmony. It involves systematically tuning into something, letting go of your mind, and then seeing if you are on the same frequency. This, by the way, is one way that ancient people taught themselves to live in harmony with nature. They said that each place had a special spirit of place. Sometimes people were sent out to meditate on a place. If they were quiet and could let go of their own minds, then—if they were lucky—a song or chant would come to them. These were the songs and chants of the ancient religions which worshiped the earth, sung in celebration of the living planet. They are part of the Earth because they are a deeply natural part of people.

It would be fascinating to do some more in-depth studies with people entering the program and exiting from it. How have they changed by being on the Institute? What have they lost and what have they gained? But as I say this, I find a conflict arising within myself. The mechanical, scientific part of me says that this is precisely what is needed here. I should be providing you the reader with some stronger quantifiable data to substantiate my impressions. They did invite me to spend nearly two weeks in Maine and paid me to come there—how can I be objective, unless I have quantifiable data? The other part of me says that the problem is with my methods. How can you really measure how alive a person is? Where is a measuring device or a control for an individual's life? How can you really know where a person would have been, had s/he done something other than what s/he did? That, it seems, is a limit to scientific methodology in this kind of situation. Beyond this, how can you measure prejudice against nature, or congruence with nature? The fact that I can't come up with a measure now doesn't mean these things don't exist. It simply says that I may not be able to measure them at this time. Maybe they must be measured against the health of the ecosystem. On this I can agree.

This was the type of discussion which we got into as we wrapped up the environmental education definition.

Conventional environmental education research deals with acquiring knowledge or applying it to concepts. It involves using paper-and-pencil questionnaires which can give us quantifiable data. These methods can help with some things; but if we define environmental education as a process of approaching person-planet congruency with the planet seen as a living being, then this may call for a whole new approach to research. When we really take the time to define what we are studying, then it seems that research works best. I could give people a paper-and-pencil questionnaire and ask them to define their concept of the earth. If they wrote down that the earth was a living being and turned in the papers, could I say that they really understood this to be what they believed? Of course I could not, and this is the problem with a lot of educational research. I believe that we need new methods to find out what we are teaching or, more importantly, what is happening to students as they go through an educational process. You can test to see if something is memorized, but what else is going on in the class? In many cases, what is learned outside of the formal class lessons is more important to the life process than what is supposed to be happening in the class. When I went to school, through the windows of my classroom, I looked out onto the Detroit River. I spent a good deal of time watching the migratory flights of ducks as they flew up and down the river. When I could, I would draw pictures of them regardless of what was going on in the class. Somehow it was where my mind wanted to go. In schools people learn about relationships—they learn how to relate to authority figures, how to make friends, even how to fall in love. No one directly teaches these things, but they happen. Conventional schools also teach that the important things happen inside buildings. Athletics is what happens outside that is educational. The subtle messages are that education takes place best when it can be structured, segmented, and presented in abstraction. This leaves us learning to lock ourselves up in school buildings inside of our heads and to treat the rest of the world this way, if we are *good students.* Quite frankly, our schools are prejudiced against human nature as well as nature. Thinking and memorizing is what counts, or at least along with learning to sit still, is what gets us good grades. This mode of education prepares us to enter into a society run just like a school. It may be contrary to human nature and to nature, but it is the foundation of the world we know today. The problem is that ours is an educational system which fragments us and asks us to become mechanical so we can more easily exist in a mechanical world.

Since Earth Day 1970, the problems of the environment have been problems that we can talk about openly. As each new door is opened to subjects that used to be taboo, we approach a more holistic education. But talking about something is very different than doing it. This, I think, may be the strongest function of the Audubon Expedition Institute. In a variety of settings across the country, they holistically study firsthand experiences which later may become extrapolated subjects like anthropology, archaeology, sociology, and natural sciences. This first-hand exposure to people and places makes subjects in books become real. They become experiences. It affords students the chance to do reality testing on what books say, and discussions about reality, especially related to environmental matters, are an important part of the Expedition. The students are encouraged to question anything, and as I learned first-hand, they do so. This aspect of the program helps people become autonomous thinkers, getting away from the weakness inherent in traditional education of relying upon rote memorization of material from authoritarian people. Authority figures have some special expertise on a subject. Because of the time and energy invested in developing this expertise, they sometimes command respect, although you may not agree with them. If you have to agree with them, then they become authoritarian, depending upon the amount of freedom you have to choose to agree or not. Authoritarianism arises out of coercion. "If you don't agree with me, then I'll flunk you," is an authoritarian position. On the Expedition, students are exposed to a whole host of authority figures. They experience them and then go away to process the experience. This is the essence of real learning.

Decision-making on the Expedition takes place on a consensus basis after having made an initial agreement to be a part of the group. Each prospective student for the program is interviewed by staff members, who explore questions of willingness to live in a group setting as well as commitment to the mission of the Institute of improving person-planet congruency. In choosing to join the Expedition Institute, a person needs to be aware of the decision s/he is making. You will be sleeping in a tent in all kinds of weather for up to two years—in some cases more, depending on the type of program you become involved with. You may not shower regularly. You often will be without electricity and hot running water. You will travel in a yellow school bus and become a member of a small community. While in this co-ed community, you will be asked not to pair up with members of the same or opposite sex for special relationships. (Experience of 15 years has shown that pairing up while on the Expedition can be very disruptive to the group process. Upon completion of their experience with the Institute, some students have become lifelong friends, while others even have married.)

While the Institute is travelling around the country, they continually examine their own behavior to see if they can reduce environmental impact. In recent years they have decided not to use paper towels, and kleenex has been replaced with handkerchiefs. During the past year clothing bearing advertising for various products which might harm the environment was questioned and discussed. Decisions like these, as well as the resulting minimal contact with radio, television, and news media tend to restore defining reality to being a personal matter.

All too often today, we become caught up in what might be called *well-informed futility*. We read all the papers and magazines, listen to the radio and watch television constantly. Although we may be well-informed, the general scope of the material may be way beyond what we feel to be in our realm of influence. We tend to let others define reality for us and lose our individual power and sense of defining reality. It's a little like people living in Los Angeles turning on the television to find out how bad the smog is, when they could just as easily step outside and see and smell it for themselves. It is true that we must recognize that we are a global society, as Buckminster Fuller has pointed out. Yet, at the same time we must question, "Just what is the nature of that global society? How does it relate to my personal world needs for my society to be defined personally?"

Recent research in the holistic health field seems to give some important clues as to the importance of personal autonomy. Studies of the personality types of people who tend to contract various kinds of cancer suggest that people who have a strong external environmental dependency for their orientation are more likely to contract cancer than those who have greater personal autonomy.[4] People who live their lives in front of television sets, letting the news commentators define reality for them, may have less awareness of what is actually going on within them. This is not to say that the television set is an evil instrument, but rather that the dependency on one may be a problem. Psychology has tended to define healthiness in terms of normality, healthy behavior most frequently falling within a bell-shaped curve of frequency. The problem with this definition is that normality may not be directly related to healthiness. If everyone in a culture is neurotic, then normality in that culture would be to be neurotic. People who did not have neuroses would be viewed as abnormal. It follows that to become a healthier society, we must move away from normality as a definition of mental health and move toward one which involves personal responsibility, lack of stress, and self-reliance. Innovation comes from people who are willing to take risks and be creative—people who are willing to deviate from the norms. The interchange between Mike and Susan Gould, the literary agent, sounds like an exchange between someone who defines normality (reality) according to statistics and someone who defines reality (normality) in terms of health and wholeness. By stripping away many of the conveniences of modern society which Mike suggests create a *technological euphoria,* people on the Institute are increasingly faced with themselves. They learn that the world does not fall apart if they don't have a television set. They learn that viable community life is the basic product of a healthy human society. They also learn that resource consumption decreases in a healthy community setting as people feel more, share more, and cooperate more.

Mike's section on competition was provocative to me. I thought back to a special conference in 1976. In Eugene, Oregon, the track and field capital of the world, we sponsored a week-long seminar entitled *Exploring the Human Potential Through Physical Activity: Mind, Body and Spirit.* The purpose of this week-long event was to bring together world-class athletes with humanists to explore the human potential as expressed through sports. Many of the participating athletes were from the Professional Track Association. Throughout the conference, competition was a topic often talked about. What seemed important was how these people defined competition. They talked about competition as being a way to push themselves to perform better. They said that being with people who could really challenge them, other people of world-class status, gave them more incentive to excel. They did not want to beat other people so much as to perform at their highest level of capability. This attitude seems to be the crucial difference between healthy competition and sick competition. Sick competitors are out to beat other people. They usually have a low self-identity and are dependent upon winning for self-esteem. Healthy competitors want to perform at their best, and hopefully when that happens they will win. Mike might say that they were agreeing to create stress cooperatively in order to achieve higher performance. Here again the culprit is fear. You have to believe in yourself to do well at whatever you do. If you have high self-esteem, then you tend to evaluate your performance more on your own standards than you do in relation to other people. Where I saw this happening on the Expedition was in their music. Many of them played instruments like banjos, guitars, mandolins and fiddles. They did not try to hold contests as to be able to see who was best. Instead they worked at everybody's developing enough proficiency to be good enough to play together in tune. This way they could have a band, play music for local country dances, and even hold concerts. The Institute band has played at the festivals held at the sloop *Clearwater* and directed by Pete Seeger, as well as at Saturday night dances in Lubec, Maine. Once they were in Arizona and met a group of Tarahumara Indians from Mexico who showed them their folk dances. In return the Institute band played folk songs for them and demonstrated the contra and Morris dances of New England. In this case neither group was trying to compete for the best dances or singing. Instead they each tried to do their best to show the essence of their culture and skills. This is the kind of competition we need to promote. We need to help people recapture their self-worth which makes it possible to express themselves without worrying about winning or losing. In such cases every person who participates becomes a winner. The whole idea of healthy versus unhealthy competition needs a good deal more study, especially in a world with a global society. We must learn to get along with each other, share our resources and conserve, and we must also find ways to encourage the expression of the fullest human potential.

It seems as though we have gone past the point in time when we can afford to continually segment people through education. In a few short years, many mechanical jobs in industry will probably be done by computers or robots. A few years ago, futurists from some camps were predicting that this would create a leisure society where *killing time* would be the biggest problem. This scenario might be true for some small percentage of the population, but I think that with the development of automation and the new technologies of the cybernetic revolution, we will cease to need to educate people to work mechanically. Just stand outside a major auto manufacturing plant some afternoon as the day shift gets off work. As the plant gates open to the sound of a horn, a swarm of men and women will come running out to get in their cars and drive away as fast as possible. They will be back the next day, but in the meanwhile will have done as much as possible to combat the stress associated with trying to be part of a machine.

As these kinds of assembly line and numerical computation jobs become taken over by real machines, the issue will not be leisure time, I think, but meaning in life. Here we run into a potential hotbed argument when it comes to environmental education, because this looking for meaning in life starts to sound like religion rather than education. Traditionally finding meaning in life has been the job of either philosophy or religion. Both tend to have difficulty actually helping people find meaning in life, because these days neither are experiential. Instead, for both religion and philosophy you are usually asked to sit quietly inside a building and memorize what an authority figure tells you. In religion, as in philosophy, the process of developing a statement or a ritual often carries more learning than does its study, unless the study can be done experientially.

In the last decade or so, it has become very popular to offer values clarification exercises in some schools and even in some environmental education programs.[5] Rather than telling people what to believe, values clarification strategies seek to help people get in touch with what they really feel. Done correctly, these exercises can be very powerful. Abraham Maslow hypothesized that real values were instinctoid—an inherited basic element of human nature.[6] He said that he felt our values were the quiet inner voices which gave meaning to our lives, which gave our lives purpose and direction. Education, he argued, tended to cover over these basic human values of concern for life, reverence for nature, love and truth, and teach us to behave according to cultural values enforced by outside authorities. Such people work well as automatons, but they don't feel easily. Deprive people of their automaton jobs and what we will be left with will be a crisis in meaninglessness.

The truth in Maslow's words may be seen in the increasing number of people who experiment with psychoactive drugs, meditation, and eastern religions. These methods of self-exploration can lead a person down some fascinating paths, yet they have problems. Little is known about the many realms of consciousness which lie beyond what we in the West call *normal waking reality,* yet most certainly there are many worlds close by which can be reached with remarkable ease. The problem with most experiential religions is that they were designed for another time and for another culture. It is the rare person from western society who can become a true and total aspirant of a traditional mystical religion. Yet it seems imperative that if we are going to offer a truly holistic environmental education, it must deal with mind, body, and spirit.

However, the great Indian mystic Ramarakrishna once observed: "There are two paths to God, one is the way of structure and the other is the way of structurelessness. They are both equally valid." The challenge of finding a structured way to unite spirit with nature resides with our churches. The way of structurelessness can be seen in the works of Emerson, Thoreau, John Muir, and Aldo Leopold. These are people who have been able to learn to communicate with nature and let nature speak through them. They all seem to have been able to do this by being able to separate themselves from normal culture and study nature first-hand for a time. Whether sitting beside Walden Pond, or clinging to the top of a swaying evergreen tree in the height of a thunderstorm as John Muir was sometimes known to do, these people seemed to be able to approach person-planet congruency much closer than most others. Their methods were not like many of the people of Earth—the Indians, the Eskimos, the Bedouins, the Bushmen or the Polynesians. In most cultures that live close to the earth, people are taught to increase their consciousness to be able to blend and flow more closely with nature on an emotional and institutional level.[7] Today when you see the hula danced in a hotel in downtown Honolulu, you may not always see the dance as it was in the beginning. Originally to dance the hula, young women and men would go off to the hills and were told to sit in silence and study the movement of a single tree in the wind. They were to remain there until they could merge with the tree and perfectly express its flow and grace in the wind. Then they could go down to the village and learn the technique and form.

It seems to me that, in a way, the Expedition Institute uses an updated version of a very old practice of learning to blend your mind with nature by first-hand exposure to wild places. Strip away your attachments to modern culture and you will find the kind of interpersonal encounters Mike describes happening. People cannot avoid each other. People cannot avoid nature. Often, in such situations, people begin to cry for reasons they cannot rationally explain. This is not insanity. Instead it is the voice of sanity itself speaking through and expressing a sob of relief at the barriers at last being broken down. I am coming to believe that the mind wants to unite with nature, within and without. This is a natural tendency. It may not always result in inceased performance score on the SAT test or on an IQ test, but it most certainly leads to a sense of personal health.

Perhaps one reason why we don't have more Thoreaus, John Muirs, or Aldo Leopolds is that these people were brave pioneers who were willing to spend a good deal of time alone by themselves in the woods. We don't have the time to have everyone go off to a Walden Pond, and a lot of us wouldn't do it. We need to find ways to connect people with nature in groups. The Audubon Expedition Institute is exploring ways of doing this; and in my view, they are achieving some success, a lot more success than many other environmental education programs or educational programs in general.

The planet is calling to us with its voices of nature saying we must adapt our ways towards learning to live in harmony with nature. We should be listening to these voices with our hearts and minds and finding ways to hear them more clearly. We need to be finding new ways to learn about how to live, before we can expect to truly grasp abstractions. Abstractions unconnected to anything run the risk of being schizophrenic. More than one educational critic these days has suggested that our schools tend to program people to be schizophrenic.

Now more than ever we need programs like the Audubon Expedition Institute. By occasionally stepping outside the confines of contemporary society and getting down to the basics of what life is all about, they enable us to tap into our own creativity and create new ways of living appropriate to the times and the planet. For many reasons it seems to me that the Audubon Expedition Institute is doing precisely this. I look to them for leadership in charting new directions for living in harmony with nature. Rather than being a fixed process, each Expedition is a new exploration of the relationship between people and nature. This definition, I think, marks the beginning of a whole series of important ideas, concepts and products which will be coming from the Expedition Institute.

It seems important to realize that the number of people who can go through the Audubon Society Expedition Institute never will be very large, even if a fleet of 20 buses and guides could be fielded. But some of the graduates will go out and start similar programs on their own and at colleges and universities. And without an Institute's pioneering experience or something comparable, I'm not sure that many college faculties will adopt programs like this, especially with today's economic constraints on college resources.

The graduates of the Institute should become leaders in the environmental education field. The challenge for them will be to find new ways to share with others experiences which will increase person-planet congruency. Can you give people an experience in a week-end or week-long seminar that will increase person-planet congruency? There are a number of programs in existence now which teach outdoor skills like climbing, hiking, and canoeing. What's the difference between them and the Institute? The group process is critical. Creating a viable community is a primary goal for the Institute. The focus is on harmonizing with nature and less on developing recreational skills.

Somehow we must find continuing ways to offer environmental education experiences that will increase person-planet congruency without the need for a solid length of time as long as the Expedition Institute. Too many people today fear nature. They build resort cabins in swampy areas and then get angry at the mosquitos. This leads them to resort either to spraying the area with pesticides or to draining the swamp. Either remedy causes more problems than the recreationists had bargained for. If there is an attractive lake nearby, the pesticides may hurt the fishing. Draining the swamp may remove breeding grounds for some fish which probably ate the mosquito larvae in the first place.

What does it take to develop a mentality which respects the limitations set upon us by nature? It seems like the idea of *false euphoria* that Mike talks about may be important here. Euphoria is an internal feeling. It can be triggered by external factors, but the feeling is internal. Could we help people learn to feel good, relax, and enjoy themselves as a way to decrease environmental pressures? Recent research on prenatal development may have some relevance, supporting some of Mike's concept of the womb or *womb-state*. Stanislav Grof, M.D., a pioneer in consciousness research, has gathered extensive data from depth psychotherapy to show that people can recall what it was like to be in their mothers' wombs.[8] Conversations between parents can be recalled, and actual experiences of being inside of a womb is vividly recalled by some people. The womb experience is one of security and soothing bliss, unless of course some trauma exists, in which case the trauma becomes the early environment of pre-birth. It has been suggested that, after birth, all people have a yearning to return to that state of bliss or euphoria that they once experienced in the womb. Meditation is supposed to be one way to regain contact with that sense of bliss. I believe that another way is to be immersed in a sustaining natural environment, free from personal threat and discomfort. This new line of consciousness research may give support to the claims of many environmentalists that we need natural areas in order to survive.

On the Expedition Institute apparently, as people let go of their cultural feelings, they become more and more capable of experiencing this state of natural euphoria in natural environmental settings. In her classical research on ecstatic experiences, psychologist Margarantha Lanski has found that natural environmental settings, such as water, wilderness or mountains, are the most frequently mentioned *triggers* of ecstacy—*triggers* are those environmental factors which seem to stimulate the necessary emotional energy to allow an ecstatic experience to take place.[9]

Until the advent of modern humanistic and transpersonal psychology, there was very little systematic research on altered states of consciousness and their relationship to human health, except for the pioneering work of William James and Carl Jung. Generally such experiences were considered abnormal and perhaps pathological, or at least the domain of religion rather than of psychology. Now, however, we know that such experiences are rather very normal, and in fact their regularity in a person's life may have some relationship to their level of self-actualization. In his research on self-actualization, Abraham Maslow found that self-actualizing people reported an increasing frequency of peak or ecstatic experiences as they approached greater levels of self-actualization.[10]

The womb-bliss of being at one with nature is one type of ecstatic experience which nature seems to be able to facilitate quite well. The archetypal symbols of nature in forests, caves, and natural settings usually are associated with healthiness and, in fact, with transformation when seen as emerging from a womb of the *earth mother*. As we explore the realm of consciousness, it becomes apparent that there are many different types of ecstatic or peak experiences. The energy-charged feelings of John Muir clearly indicate his *mystical* nature and also suggest the power of certain places to evoke feelings of wonder, transcendence, awe and rapture. My own research in this area, gathered from several years of interviews with people who have had mystical experiences, shows quite clearly that natural settings are powerful triggers for states of transcendence.[11] A survey of cross-cultural references to the role of nature in transcendence shows that many cultures even have special places which are supposed to have special powers to be able to evoke mystical experiences. The Sioux and other Indians claim that such places as the Black Hills of South Dakota have a special power and are called *sacred places*. Some places are said to have such power that only very strong mystics should attempt to have a vision there, for those of weak mind will instead go crazy. Mount Patrick Cougall in Wales, for example, has a legend that if you spend a night on the summit of the mountain, you will come down either a poet or a madman. That such places do have a special power to evoke mystical experiences is still open to study. That people do tend to have vivid experiences at these places is a matter of record. Let us suppose for a moment the places do have such special power. Then the value of such a natural environment is the power to evoke deep insights into human nature. Such dreams and visions are the wellspring of human creativity. Although difficult to communicate through language, they are the very basis by which we invent new ways of successful living on earth. This is another powerful argument in favor of the preservation of wilderness. It is an area which needs a good deal more research. Needless to say, I have encouraged Mike and Diana to keep careful records of the strong, unplanned experiences which people report having in special places. Through such records, the Expedition Institute may be able to help considerably with our understanding of the relationship between experience and place. We may well find that the reason, or at least one reason, why *mother nature* is so important to us is that she has the power to help us re-create ourselves. Nature may be seen as an amplifier of human states of re-creation.

In talking about this subject of the variety of experiences of person-planet congruency, we run up against the problem stated in Working Assumption No. 5: our lack of an adequate language to explain the necessary states. To this end, Mike's experience with the CNS paradigm, anti-Nature myths, and Whole-Life factor deserves widespread consideration. Cultures which conduct ceremonies and rituals and practice lifestyles which are seemingly more congruent with person-planet harmony use phrases like *earth mother* to refer to the planet. Such phrases produce images which subtly suggest aliveness and the need for a sense of respect and love for the earth as we might respect and love our own mother. Because of the mechanical nature of our culture, such images aren't normally used in our language. This may seem like a picky point, but really it is not. The structure of our sentences and the emotions and images they contain set a tone which pervades and limits our lives. In modern western culture, if we see an eagle fly by, we think of the word *eagle*. If we are more well-versed in ornithology, we might want to see what species of eagle the bird is. A natural scientist might even think in terms of Latin names as well as English. Contrast this with an American Indian who sees the same eagle.

The eagle, American Indian children are taught, symbolizes the direction of the East, which is the direction of illumination. The eagle brings the power of sight and vision into the future. An eagle also symbolizes truth which is the factor that makes vision into prophesy. Therefore, if a western scientist and an American Indian are standing side by side on the same hillside watching the same eagle, they probably are having very different experiences inside due to cultural conditioning. Generally speaking, modern western culture does not place a great deal of overt emphasis upon the symbolic feelingful importance of things. We have a rational, pragmatic culture. This no doubt is one reason why western psychology has such a hard time adequately describing intuition, which tends to operate on a feeling and symbolic level, speaking more directly to the human unconscious.

Myths, fables and fairy tales represent one way in which we do study symbolic meaning in western culture. Environmental education needs to devote a good deal more attention to myth and symbol. Some existing fairy tales contain negative environmental messages, like the *big bad wolf* in Little Red Riding Hood. Why not collect and create some myths that enhance nature? This seems like another project ideally suited for the Expedition Institute. No doubt the land-based cultures they visit do have special folk tales which aren't that well known. Collecting these and sharing them with us in song and story would be helpful towards improving person-planet congruency. They might also research other folk arts like story-telling to see how person-planet congruence is subtly communicated in other subcultures.

The overall goal of promoting person-planet congruency seems now ready to be openly discussed and explored. The idea that the earth is a living being is no longer simple fantasy. Increasingly, research from many different disciplines is coming to the conclusion that in one way or another this is so. It is the goal of the mystic, the visionary, the transcendentalist, the religious leader, and the scientist to perceive the true nature of things. For too many years, they have been at odds with each other instead of mutually supporting and confirming each other's observations. Concurrent validity, when several different scientific approaches arrive at the same conclusion, is one of the strongest tests of truth. The common perception of the earth as a living being may be one of the most powerful concepts we can have—a positive focus for our efforts to restore the many life systems of the earth to health and vitality. Seeing ourselves as a part of nature, rather than as separate from nature, is of vital importance to this process of restoration. For the health of us all, persons and planet, we need to erase our prejudice against nature and restore harmony to the world. We need to allow ourselves to be able to sense and maintain a sense of common greenness about the living system of the planet.

—*Jim Swan, Ph.D.*
International College
1983

REFERENCES

[1]SWAN, J., and WM. STAPP. *Environmental Education: Strategies Toward a More Livable Future.* New York, New York: Halstead Press, John Wiley and Sons, Inc., 1974.

[2]LINGWOOD, D. *Environmental Information-Seeking Through a Teach-In* in Swan and Stapp, *Ibid.*, pp. 174-206.

[3]SWAN, J. Response to Air Pollution: A Study of The Attitudes and Coping Styles of High School Youth, *Journal of Environment and Behavior,* vol. 2, September, 1970.

[4]GREEN, E. and A. GREEN. *Beyond Biofeedback.* New York, New York: Delacorte Press, 1978.

[5]SWAN, J. "The Formation of Environmental Values: A Social Process" in Cook and O'Hearn (eds.) *Processes for a Quality Environment.* Green Bay, Wisconsin: University of Wisconsin Press, 1971.

[6]MASLOW, A. H. *The Farther Reaches of Human Nature.* New York, New York: Viking Esalen Press, 1971.

[7]HALIFAX, J. *Shamanic Voices.* New York, New York: Dutton, 1980.

[8]GROF, S. *Realms of the Unconscious.* New York, New York: Dutton, 1976.
     (Dr. Grof is also founder of the Spiritual Emergencies Network, which is centered at the Esalen Institute in Big Sur, California.)

[9]LASKI, M. *Ecstasy.* Bloomington, Indiana: University of Indiana Press, 1964.

[10]MASLOW, A. H. *Op. Cit.*

[11]SWAN, J. Transcendental Experiences in Nature. *Theta Journal.* Society for Psychical Research, 1983.

APPENDICES

As long as man dwells in a state of pure nature (I mean pure and not coarse nature), all his being acts at once like a simple sensuous unity, like a harmonious whole. The senses and reason, the receptive faculty and the spontaneously active faculty, have not as yet been separated in their respective functions; a fortiori they are not yet in contradiction with each other. Then the feelings of man are not the formless play of chance; nor are his thoughts an empty play of the imagination, without any value. His feelings proceed from the law of necessity, his thoughts from reality. But when a man enters the state of civilization, and art has fashioned him, this sensuous harmony which was in him disappears, and henceforth he can only manifest himself as a moral unity, that is, as aspiring to unity. The harmony that existed as a fact in the former state, the harmony of feeling and thought, only exists now in an ideal state. It is no longer in him, but out of him; it is a conception of thought which he must begin by realizing in himself; it is no longer a fact, a reality of his life.

—Schiller, *Simple and Sentimental Poetry*

Every part of this earth is sacred to my people. Every shining pine needle, every sandy shore, every mist in the dark woods, every clearing and humming insect is holy in the memory and experience of my people. The sap which courses through the trees carries the memories of the red man.

I am a savage and do not understand any other way. I have seen a thousand rotting buffalos on the prairie, left by the white man who shot them from a passing train. I am a savage and I do not understand how the smoking iron horse can be more important than the buffalo that we kill only to stay alive.

The white man's dead forget the country of their birth when they go to walk among the stars. Our dead never forget this beautiful earth, for it is the mother of the red man. We are part of the earth and it is part of us. The perfumed flowers are our sisters, the deer, the horse, the great eagle, these are our brothers. The rocky crests, the juices in the meadows, the body heat of the pony, and man—all belong to the same family.

—Chief Sealth, Duamish Tribe, 1854

259

ONE HUNDRED AND ELEVENTH LEGISLATURE

Legislative Document   No. 967

S.P. 322   In Senate, March 3, 1983

Referred to the Committee on Agriculture. Sent down for concurrence and ordered printed.

JOY J. O'BRIEN, Secretary of the Senate

Presented by Senator Brown of Washington.
Cosponsors: Senator Wood of York, Representative Crouse of Washburn and Representative Michael of Auburn.

STATE OF MAINE

IN THE YEAR OF OUR LORD
NINETEEN HUNDRED AND EIGHTY-THREE

AN ACT to Establish a Numerical Whole
Life Factor to be Placed Voluntarily on Food
Items Sold at the Retail Level in Maine.

Be it enacted by the People of the State of Maine as follows:

Sec. 1. 22 MRSA §2152, sub-§7-B is enacted to read:

7-B. Numerical whole life factor. "Numerical whole life factor" means a single number assigned to a food product which reflects the labor intensity, energy consumption and resource use and depletion involved in harvesting, processing, packaging, transporting and distributing a food item to retail food buyers. The whole life factor, expressed as a 1-100% index, measures the relative costs to the ecosystem in producing various food products.

Appendix A

THINKING GLOBALLY: THE WHOLE LIFE FACTOR
If there is one way better than another, it is the way of Nature—Aristotle

LD 1983: An Act to Apply a Numerical *Whole Life Factor* to All Products and Services in Washington County, Maine.

Support materials for a 1983 Legislative Document of the State of Maine:
(See "What's In This Package," p. 226)

Introduction and Overview:

Within the last several decades, there has been an increasing amount of concern expressed about the quality of life on earth as well as the survival of the planet's living ecosystem. Research reports, such as The Club of Rome Study, *The Limits to Growth,* and the recent U.S. Government report on alternative futures for the year 2000 and beyond, agree that the planet's serious problems of environmental pollution, resource depletion, and energy consumption could worsen to the point of threatening the survival of many species on earth including human beings. In light of these findings, a number of efforts have been made to address these problems. New laws, new publications, new organizations, new lifestyles, and new products are among the many things that have been spawned from a growing concern for the earth and the preservation of life.

The environmental education field and movement, while having its roots in outdoor and conservation education, has also grown significantly during the last few years. The basic attempt has been to familiarize the adult and youth populations of the United States and the world to our environmental problems and present some ways to become involved in resolving them. As the environmental movement and environmental education have grown from a response to a set of problems to a more holistic overall view of the relationship between the human race and the planet, it has become increasingly clear that resolving environmental problems will require a restructuring of many elements of consciousness and feelings as well as behavior. A careful study of this problem, for example, reveals the distressing situation that we do not even have a good working language to adequately describe the many dimensions of the person-planet relationship. The recent introductions to our working language of the words *ecology, ecosystem,* and *synergy* have made great contributions, and are examples of the need to further develop a new language to adequately articulate this problem. We must move from our conventional linear, rational, analytical, deductive thinking to also embrace a holistic systems approach to thinking in order to account for the working nature of the planet, people, and the relationships of the many individual elements which support the existence of life itself. One way to a full approach expressing this relationship is to see the earth as a living being. This is a concept which has been held by many, if not most cultures and has been proposed recently by western scientists as well. Using the concept of the earth being a living organism, one can begin to talk about the relationship between people and the earth in terms of harmony, or congruency, implying that living in harmony with nature or person-planet congruence results in health for both the planet at large and the individual species including humankind. This concept is exciting. It is being used on already existing groups to produce holistic relationships that provide increased support for the planet and people.

Integrating this living earth concept into our culture must be done from a variety of perspectives to meet the multifaceted needs of human beings, as well as to express their uniqueness in relation to the life systems of the planet. Attitudes, values, skills, perceptions, feelings, knowledge, and behavior must be developed to create a new world view or *paradigm* which totally reflects the *earth as a living being* concept. We must develop a practical new wisdom of guidance to enable the continued integration of the human species on local and global levels as a part of the planet.

An integral element of this overall paradigm creation is the synthesis of a new language and with it a continual process of people-planet education. The purpose of this proposal is to develop a special working seminar to refine and develop what may well be a very important element of this new language. The concept proposed to be refined and developed is what we are calling the *Whole Life factor.* The Whole Life factor is defined as a numerical value, which for the purposes of this proposal may be assigned to a product or service, and which accurately describes the degree of local harmony or congruence that the product or service has with the preservation and enhancement of the life processes of the living earth, including the survival of the human race.

Implications:

Although average consumers today are environmentally concerned, they are often bewildered by trying to make choices consistent with the idea of supporting the environment, i.e., congruent with the Whole Life factor concept. Though there is an overall awareness of environmental problems, translating this awareness and concern into action is often impossible. The almost overwhelming array of political and economic pressures, facts and figures, viewpoints, and technical jargon involved can easily lead to the creation of a condition which might best be called "well-informed futility." There is a desperate need to be able to easily comprehend the history and nature of person-planet congruency, and to translate feelings of concern into effective action. For many people, to become involved leads to an enervating problem of increasing complexity, which in the long run leads to anxious confusion and a helpless apathetic sense of decreased concern and commitment to improve the person-planet relationship.

Within the last two decades, there has been a tremendous increase in the mass education of consumers. Magazines such as *Consumer Reports* have addressed the quality of products and services, as have more detailed studies of various products and services by organizations such as Ralph Nader's task force. The results often have been significant; and at the same time, no one has taken this additional data and sought to translate it into an environmental health language which is easily communicated and understood. We propose that one remedy for this problem is to create a single numerical factor which expresses person-planet congruency. We call this the *Whole Life factor* (WLF). Using this simple concept as a core element in communicating the living earth philosophy, it seems likely that magazines, newsletters, and other publications would effect and communicate this concept and its implications to the general public. Since the more simple expression of a concept is the better one, we suggest a single number, expressed as *WLF 0-100% congruency* to help our language, thinking, and actions become more aligned with nature. Such a simple concept could be taught to young children as well as to adults and could be communicated in many ways, including being placed on the label of a product or service, just like the ingredients in food that are now required on all packaging. It could be used as a theme for creating festivals to celebrate the concept of the living earth, as well as to promote products and services which have favorable WLF values.

The Whole Life Factor:

Modern western culture has a strong scientific and technological focus which includes a trust in numerical symbols and values. Numerical values are often used to express complicated concepts, such as the acidity of water, the concentration of one substance in another, school grades, or even the attitudinal preferences of judges about the performance of athletes. What is important is the practical introduction of the whole life

value into the daily lives of people, and the articulation of the derivation of
the value, as well as the validity of the method to accurately assess what it is
proposed to assess.

In developing a Whole Life factor numerical value, we feel it would be
important to consider such factors as: energy consumption, geographical
locality, labor intensity, resource use and consumption, transportation,
toxicity, packaging, exploitation, and likely environmental consequences
of usage. These factors all could be expressed as congruency formulas
leading to numerical values which could be collectively integrated and
expressed as a single number. It would be similar to a numerical sliding
scale environmental impact statement for public guidance.

To be a truly holistic measurement, we also suggest that a subjective
evaluation of person-planet congruency be included in the WLF
calculation. This might be done by the consensus opinion of a panel of
expert judges, who are comfortable with the belief that the planet is a living
organism. They would look at such aspects of the product or service as
design efficacy (as per appropriate technology), advertising as congruent
with the WLF spirit, endurance, recyclability, comparative necessity of the
product or service, attitudinal impact, present and projected state of the
environment, and capacity to enhance human health and well-being as well
as that of the overall life systems of the planet. Such subjective evaluations
are commonly done in athletic competition as well as assessing the quality
of foods, like wine-tasting, etc. Hearings and testimony from all parties
involved could be integrated into the panel's decisions.

These two figures could then be combined into a single number, the
Whole Life factor. When it was assigned, cooperating businesses would be
legally bound to maintain the quality control and processes under which it
was derived, or notify the committee if changes occur.

The Research and Development Process for the Whole Life Factor:

Because the Whole Life factor is such an important and practical
concept, it is proposed that it be developed in consultation with some of the
most articulate and knowledgeable people who have special expertise
concerning person-planet congruency. These experts would be called
together at a retreat center for 3-5 days in closed session to refine and
develop this concept. Also in attendance would be experts in computer
simulations and communications experts who would develop programs to
compute and express the Whole Life factor.

Upon the acceptance of an invitation to such a conference each expert
would receive a description of the concept and purpose of the meeting.
Additionally, phone conversations would be placed to insure
understanding. Then each person would be asked to prepare a short

statement of what elements they feel should be included in the creation of the Whole Life factor. These statements would be reviewed, synthesized, and returned to the experts, perhaps with additional questions or suggestions. This would begin to develop a working knowledge of the group consensus before actually meeting. Then when participants did meet, discussion would be more directed and problem areas more easily identified. For example, at the retreat, the objective and subjective measures might be discussed in smaller groups and then shared with each other. The computer specialists would take the concepts generated and translate them into programs to allow for possible pilot testing of some items.

As a follow-through, some individuals might be sought out to advise on the refinement of some aspects of the concept as well as its use and presentation. Legal advice on the use and presentation of the concept would be essential before release to the general public.

Having achieved the operationalization of the Whole Life factor, the product and services research could begin, as well as the creation of ways to publicize the results and explain the derivations of the WL factors as they were assigned, such as a magazine, newsletter, or radio and television programs. Self-perpetuating funding could be generated from grants from industries and services which have or desire Whole Life factor evaluations. Festivals to feature and support positive Whole Life factor products and services could also then be planned and later carried out.

Timeline:

With adequate funding, we suggest that a Whole Life factor and a beginning of a publication of products and services rated for their Whole Life factor could be accomplished within six to eight months following the start of the project. Funding for this project might be done in three phases.

1. *Phase One—Conference Planning*—During this phase, experts would be contacted as consultants, and materials would be prepared to orient them to the project. The initial questioning and concept formation could be done, and the size of the seminar could be decided upon, as well as the location and dates. Most of this could be done on the phone and with a small number of active researchers.

2. *Phase Two—Conducting the Conference and Developing the Terms for Mass Communications*—Here a larger staff would be helpful as well as a determination of the maximum size of the group. It seems likely that a small working group would be preferable to a larger group—with perhaps 8-12 experts as a size range for maximum effectiveness. Budgeting here would include travel, room and board, honoraria, computer utilization, and related resources.

3. *Phase Three—Developing a Research Team and Communications Team for the Whole Life factor*—To apply the WLF to many products and services will require a staff of researchers and experts to evaluate products and services subjectively. Also needed would be communicators to translate the research into ways for people to understand and apply it.

Appendix B

Outlines of Some Courses of Study That Investigate Aspects of Prejudice Against Nature

The full intent of the courses that follow is reached when the subject areas that they cover is learned experientially, on a thought, feeling, physical, and sensing level. The courses represent a challenge to the creativity of student and teacher alike. The goal is to design or recapture person-planet experiences that illuminate each of the topics within the personal framework of the course participants. It is suggested that if a specific experience is missing or needs re-examination, an exercise or expedition to gain that experience be designed and actualized.

A few books are suggested to accompany each course, but because of our learn-from-the-environment approach, our booklist is incomplete. Part of our graduate school coursework is to find and evaluate books that provide further appropriate information, or, as a project, to write a class book that provides information based upon the experiences of the students and staff. The reader can do the same.

The Expedition Institute would be interested in receiving and reviewing materials that are used for or developed from these courses and assist in their publication.

Course Goals & Rationale by Trudy Phillips

In order for us to establish a healthy relationship between ourselves and the natural world, it is imperative that on thought, feeling, and action levels we recognize that we are prejudiced against nature and that through re-education we can learn to relate to the planet as an essential, valuable living organism. To do this, it is necessary that we learn how to contact our unconscious feelings from and about the natural world, learning how to constructively act off of these feelings by fully integrating them with rational and cognitive understandings. We must:

1) Recognize that the Earth communicates through natural feelings and examine the planetary origins and survival value of those feelings.

2) Learn to design and establish environments which will support these feelings so that they can safely surface and gain mutual support.
3) Learn how and why these natural feelings have been repressed.
4) Learn the reasons and processes behind which self-preserving, earth-harmonious natural feelings become disharmonious cultural feelings.
5) Recognize the conflict between natural and cultural feelings and locate the source.
6) Develop and reinforce behaviors and settings congruent with maintaining the earth as a living organism.

Objectives:

1) To provide experiences through which the students can recognize and understand the relationship between themselves, their natural feelings, and self-preservation.
2) To understand the process and effects of prejudicial attitudes and behavior.
3) To introduce students to the role of the media (magazines, newspapers, T.V., and radio) in perpetuating their separation from nature.
4) To have students explore some of the origins of prejudice against nature (i.e. womb euphoria vs. Nature's fluctuations).
   a) To explore the historical role of European ethics (i.e. puritanism, protestantism).
   b) To explore the influence of our cultural conditioning and culture as society's statement about nature.
   c) To explore Native American relationships with the natural world.
   d) To understand the nature of language and its relationship to prejudice against nature (PAN).
5) To investigate ways in which we daily separate ourselves from the natural world.
6) To introduce field experiences during which the students explore their feelings about, and relationship to, the natural world through: a) a 3-5 day expedition experience involving extended exposure to the natural world, and b) solo experiences.
7) To create an awareness among students of the process of prejudice: a) cultural myths, b) traditions, c) literature, d) music.
8) To expose students to specific environmental issues, focusing on the relationship of the PAN concept to the issue.

9) To have each student design an experience for contacting PAN feelings. An experience they would like to use with their future students.

10) To explore ideas which could be helpful in changing the course of PAN.

11) To have students gain a comfort level with nature through back-packing; to have experienced backpackers gain a new perspective on a familiar activity.

12) To experience feelings of gratification while learning to enjoy Nature's tension and stress. (i.e. snow, rain, heat, walking, running, crawling, swimming, crafts)

13) To explore the role and effective use of the following concepts: confrontation, the Whole Life Factor (WLF), the CNS paradigm, the ten ammendments, and the myths of PAN, to bring PAN to consciousness. (See Appendix A.)

14) To take a look at statements of philosophers, writers, statesmen, politicians, etc. which are PAN or confront demonstrative aspects of PAN.

15) To explore substitute euphorias and their impact upon the natural environment.

16) To have students recognize the divisionary aspects of "academic subjects" and the PAN that might be in them.

17) To recognize the full impact of our society's unwritten goal of "distance from nature is status."

18) To understand the parallels and relationship between PAN and prejudice against people.

19) To have students recognize and identify PAN thoughts, feelings, and acts in the world around them.

20) To have students begin to recognize and identify PAN thoughts and feelings in themselves.

THE PREJUDICE OF SCIENCE AND TECHNOLOGY AGAINST NATURE

MAJOR COURSE OBJECTIVES: The course objectives are for students to locate prejudice against nature in science and technology and to pursue personal and public understanding and communication of this paradigm.

COURSE CONTENT

    I. The nature of Nature:
        A. Methods of defining and determining Nature;
        B. The forms of Nature: life, death, animate, inanimate;
        C. Are people Nature?

II. The nature of science:
  A. Does science exist, what is it?
  B. Methods of defining and determining science;
  C. The values of science;
  D. The philosophy of science;
  E. Does science teach a point of view?
III. Science and natural laws:
  A. What are natural laws?
  B. Are natural laws real, imagined or imposed?
  C. Are natural laws a creation of science?
  D. Are natural laws a means of exploitation?
IV. The role of technology:
  A. What is a technology?
  B. Technology's congruency with Nature;
  C. The impact of technology upon Nature.
V. Wholeness and science:
  A. Does the whole equal its parts?
  B. Science and subject areas;
  C. How science can find wholeness.
VI. The science of psychology:
  A. The reality of emotions;
  B. Emotions in Nature;
  C. Universal emotionality;
  D. Are emotions congruent with Nature?
VII. Prejudice against Nature:
  A. Creating distance from Nature;
  B. Science creating mistrust of Nature;
  C. Technology building confidence in culture.
VIII. Subduing the prejudice in science and technology against Nature:
  A. Designing and locating situations where science and technology have a mutually beneficial relationship with Nature;
  B. Rectifying situations where science and technology are exploiting Nature.

SUGGESTED READINGS:
So Human an Animal. DUBOS, RENE.
Entropy. RIFKIN, JEREMY.
Prejudice Against Nature: A Guidebook for the Liberation of Self and Planet. COHEN, MICHAEL J.
The Deltoid Pumpkin. MCPHEE, JOHN.
Small is Beautiful. SCHUMACHER, E. A.
Animal Liberation. SINGER, PETER.

THE PSYCHOLOGICAL BASIS OF PREJUDICE AGAINST NATURE

MAJOR COURSE OBJECTIVES: The major objective of this course is to develop an awareness of the congruency between the living planet Earth and ourselves, and to discover the basis of that congruency.

COURSE CONTENT:

I. The emotionality of the living planet:
   A. The life system of the planet as an extremely ancient process;
   B. The planet's ability to globally maintain the parameters of life;
   C. Existence of communication to maintain planetary self-preservation;
   D. Tension and tension release as planetary communication.

II. Womb-earth congruency:
   A. Exploring the emotional atmosphere of womb, euphoria, and the associative process;
   B. Comparisons of the womb environment and that of the planet;
   C. The presence of self-preservation consciousness in the womb environment;
   D. The chemical dependence of life...recognition of the human womb.

III. The psychological operants within the human embryo:
   A. The ability to sense stimuli;
   B. The ability to remember feelings;
   C. The ability to respond to stimuli;
   D. The existence of memory;
   E. The feelings of dependency;
   F. The operants of associations.

IV. Emotionality and the birth process:
   A. The inception of anxiety;
   B. Emotional memories of birth;
   C. Stress, tension, recovery, and dependence;
   D. Recognition of the human mother.

V. Infancy and childhood:
   A. The pre-natal and post-natal comparisons;
   B. The fluctuations of Nature;
   C. The process of conditioning;
   D. The message of culture and technology.

VI. Language and culture:
   A. Natural world as a symbol;

   B. Symbolization of the natural world;
   C. Technology as a part of culture.
VII. Childhood emotionality within the adult:
   A. Conscious and unconscious feelings;
   B. Natural feelings and cultural feelings;
   C. Self-preservation attitudes;
   D. Cultural preservation attitudes;
   E. Feelings of independence;
   F. Recognition of Mother Earth.
VIII. The prejudiced adult:
   A. Unconscious comparisons between culture and Nature;
   B. The reinforcement of culture;
   C. The authority of the immediate environment;
   D. Dislodging the resolved emotionality of prejudice.
IX. Offsetting prejudice against Nature:
   A. Planning a personally conceived program for subduing prejudice against Nature in self;
   B. Planning a personally conceived program for subduing prejudice against Nature in other individuals;
   C. Planning a personally conceived program for subduing prejudice against Nature in institutions.

SUGGESTED READINGS:
   Symbols of Man. JUNG, CARL.
   Psychology of Consciousness. ORNSTEIN, ROBERT.
   Prejudice Against Nature: A Guidebook for the Liberation of Self and Planet. COHEN, MICHAEL J.
   Night Country. EISELEY, LOREN.
   The Anatomy of Human Destruction. FROMM, ERIC.

THE COMMUNICATION OF PREJUDICE AGAINST NATURE

MAJOR COURSE OBJECTIVES: The objectives of the course are to give an overview of the tension-release process (T-R) and the methods of T-R used by the planet to maintain and preserve its life systems.

COURSE CONTENT

   I. Tension-tension release (T-R) communication:
      A. Experiential communication;

B. Communication through the build up and release of energy bonds;
C. As a function of energy;
D. Insinuates a universal consciousness;
E. Its purpose may be the self-preservation of life.
II. T-R global communication:
  A. Maintenance of appropriate parameters—
    1. Temperature,
    2. Salinity,
    3. Oxygen and carbon dioxide balance,
    4. Atmospheric gas ratios,
    5. Minerals.
  B. Geological communication;
  C. Seasonal communication.
III. T-R in plant and animal systems:
  A. Genetic code energy release;
  B. Irritability of protoplasm;
  C. Territorialism;
  D. Aggression;
  E. Anxiety;
  F. Natural needs and niches;
  G. Seasonal changes and signals;
  H. Cyclic release.
IV. Human communication and T-R:
  A. Self-preservation feelings;
  B. Natural feelings;
  C. Non-feelingful communication of some aspects of culture;
  D. Language, derivations, symbols, images, advertising and media;
  E. Body language;
  F. Scientific readout;
  G. The arts: music, dance, and drawing;
  H. Cultural feelings.
V. The utilization of Nature, fears, and anxiety by culture:
  A. Mythology of fear—
    1. Childhood,
    2. Media.
  B. Rites of passage—
    1. Biological and cultural milestones.
  C. Working adult—

VI. T-R and prejudice:
   A. The energy impact difference between cultural and natural communication;
   B. The effect of artificial T-R through technology and culture;
   C. The fear of natural feelings;
   D. Utilization of fear and anxiety about Nature by culture;
   E. The absence of T-R that is associated with self-preservation.
VII. Offsetting psychological prejudice against Nature:
   A. Personal histories of the communication of prejudice against Nature;
   B. Discovering present reinforcement operants;
   C. Planning personal programs to offset prejudice against Nature.

SUGGESTED READINGS:
   Little Green Book. LOBELL, JOHN.
   Prejudice Against Nature: A Guidebook for the Liberation of Self and Planet. COHEN, MICHAEL J.
   Civilization and its Discontents. FREUD, SIGMUND.
   Equilibrium of Survival, Songs of Humanity and Nature. FOLKWAYS RECORDS.

APPENDIX C
General Distinctions Between Traditional Education and Audubon Expedition Education

July, 1980, Washington, D.C. (UPI)—The federal government's most comprehensive study of the future laid down a grim warning of a world that will be increasingly crowded, hungry, polluted and unstable by the year 2000. The massive study **Global 2000** states, "world population growth, the degradation of the earth's natural resource base, and the spread of environmental pollution collectively threaten the welfare of mankind." The study, more than three years in the making, says even soil to grow food will be in short supply, as the earth is washed away or poisoned. Pure water, heavily polluted by pesticides used to increase food production, will be scarce. Coastal waters, the best source of fish, will be threatened by increasing pollution. The air will be fouler and the atmosphere will have less oxygen because of the disappearance of the large forest areas and the consumption of non-renewable resources such as coal and oil. Even the renewable resources such as trees, water and hundreds of species of birds and animals will disappear—unless the threat is recognized and action taken.

The Audubon Expedition Institute is a system of experiential education which is designed to teach the competencies and attitudes necessary to counteract the projections of **Global 2000.** In general terms, what distinguishes AEI from traditional education?

Typical Traditional Education	Audubon Expedition Education
Memorization of facts trains the mind to think.	Attitudes and competency develop in response to today's environment.
Courses are taught from a prescribed syllabus of globally unrelated facts.	Courses reflect real-life encounters with people and places.

275

Learning is achieved almost exclusively via on-campus classrooms, printed materials, and professional educators.

An infinite range of resources involve the physical and social world.

Daily personal problems, questions and relationships are unrelated to academics.

Academics are based on the students' curiosity, feelings, and relationship to the community and environment.

The society which established traditional education is mimicked, thereby reinforcing contagious dependencies and mistakes.

Critical thinking and ongoing experience make up the learning community.

The environment is considered an exploitable natural resource.

The environment provides an integrated sustaining relationship between nature and humanity.

Students live and learn in an indoor setting, seek escapes, and become dependent upon technology.

Participants live and learn in an outdoor setting, seek encounters, and become self-reliant.

Students' goal is to memorize facts, pass tests, get good grades, and receive economically valuable degrees.

Participants learn to survive successfully in equilibrium with nature, society, and self.

Faculty members see selves primarily as transmitters of information and knowledge.

Faculty members see selves primarily as counselors and expeditors of learning.

Testing, and awarding of credits and degrees depend almost exclusively on mastery of course content.

Finding out how and what to learn are as important as cognitive learning.

Dependence on authority, conformity, and competition are cultivated through mass media, prescribed curricula, required campus residence, and compulsory classes.

A cooperative sense of community, self-direction, and autonomy is cultivated through student-planned study and survival.

The curricula is predominantly oriented toward environmentally questionable traditional disciplines.

Learning is problem-oriented, issue-oriented, and world-oriented.

Students are graduated as *finished products*.

Students become lifelong learners capable of responding through life to their own evolving needs and those of society and the environment.

Faculty live a technologically dependent lifestyle. They are not required to model or practice what they preach.

Faculty live an ecologically reasonable lifestyle year-round.

Appendix D

This article's approach to the concept that the planet is alive represents an in-vogue physical science discussion of the subject, and it is, therefore, publishable. There is no evidence that the article's line of reasoning is more accurate than the author's personal thoughts, feelings, and associations that are described on page 64, yet the latter have comparatively little value to mainstream readership because of our cultural warp towards science.

For the Love of Peat
by Mike Cohen

An editorial from *The Bangor Daily News, August 26, 1982*

If you're positive that the world is only 6,000 years old, then don't bother reading any farther. Scientist's date the fossil evidence for life on the planet to go back more than 3 billion years; they have discovered that during that period the sun grew 30 per cent hotter.

The significance of these calculations escaped my mind until I learned that even a 2 per cent plus or minus change in the earth's temperature could cause either extreme glaciation or an almost boiling hot planet like Venus. Either would terminate life.

In addition to potential temperature extremes, the planet has in its lifetime survived large meteor collisions that may have caused the formation of Hudson Bay, incredible volcanic periods, dust and atomic clouds that blackened the skies for hundreds of years, and conditions that would make the seas extremely salty the air full of methane, sulfur, ammonia, or other noxious gases, all of which would have terminated life.

Yet the fossil record is continuous; it shows that life in many forms survived through all these challenging billions of years. How could this be possible?

Presently the most plausible explanation is that the planet itself acted like a living organism which maintained the environmental parameters necessary for the continuation of life. The planet has demonstrated that it can sustain a proper global body temperature and air and water quality in the same way that our body maintains the proper temperature, fluid levels, and proportions that sustain the life of our organ systems and thereby ourselves.

To put it bluntly, there are strong indications that the planet is a living organism of which people, plants, and animals are organs, just like a living person has a living heart.

Western scientists are first discovering what other cultures have known since time immemorial. Our planet is known as Mother Earth because it is a gigantic living being that mothers us by globally and organically maintaining and regulating all of the factors that permit the planet and life to exist.

Residents of eastern Maine should be extremely interested in the specific mechanisms by which Mother Earth enables her own survival as well as ours, especially with respect to the 21 per cent oxygen content of the air we breathe (if that figure went up to 25 per cent we would be unable to extinguish forest fires).

The oxygen content of our atmosphere was not formed by plant photosynthesis alone. Most of the oxygen produced in this manner is used up in plant and animal respiration. What the planet seems to do is produce our atmospheric oxygen by burying plant carbons such as coal and peat. These carbons would otherwise combine with free oxygen and form carbon dioxide and produce the greenhouse effect. The oxygen of our atmosphere today was in ages past chemically combined with carbon in plants; by separating the carbon from the oxygen and burying the carbon, Mother Earth produced our 21 per cent oxygen atmosphere.

In addition, Mother Earth regulates the oxygen level; if the oxygen level gets too low, Mother Earth may speed up the burial of carbon, the peat and coal producing process, thus releasing more free oxygen. Vice versa, if the oxygen content is too high, the rate of peat and coal formation, may be retarded.

This process isn't a physical or mechanical act. It's an organic living regulation of the atmosphere by the planet itself. It is consciously regulated by the growth rate of peat moss and other plants. It is the same as one of us breathing faster or slower in order to maintain our proper air supply as conditions change. That's something I think about as we plan to destroy living peat bogs in order to obtain peat for energy. It appears to me that the damage will be the loss of a living heath, the loss of non-renewable peat, the loss of oxygen from burning carbon deposits Mother Earth buried thousands of years ago in order that our atmosphere might support life, and an increase in the atmosphere's carbon dioxide content.

There is another amazing facet to peat. Unusually good evidence demonstrates that living peat moss gathers moisture; by so doing it organically makes bogs and lakes that I once thought were made geologically. Peat moss may also be able to regulate the water table, and local temperature and weather conditions by regulating the process of evaporation. In the same way that through evaporation our sweat glands and pores regulate our temperature and water content, peat moss serves the life of Mother Earth and therefore ourselves. Yes, it is *just peat moss*, but it is also alive, and it is one of the organs of the global life system. Maybe that is why there are so many bogs.

Acid rain is an example of economics that hurt the living earth. We should be sure to keep the health of Mother Earth and ourselves in mind as we now roll out the invitational carpet in eastern Maine for the peat mining companies from away. Could we ever afford a workman's compensation policy that covers Mother Earth when we injure her? What is the cost of acid rain? Are economics that injure the planet wise economics?

Appendix E

Acknowledgements

Special thanks go to Selden West, Straight Bay, Frank Trocco, The John Muir Trail, Diana Cohen, Everglades, David Laing, Coffin's Neck, Serena Lockwood, Adirondack backcountry, Ben Williams, Pilgrim Beach, Dan Tishman, Old Sturbridge Village, Trudy Phillips, Sunshine Mine, Sonyanis, Uinta Primitive Area, Elijah Good and family, Havasupai Canyon, Trudy Trocco, White Camp, Alan and Debbie Furth, Oak Bluffs, Joanne Sharpe, Timber, Burlington Mall, Tweed River, Tom Wisner, Buckskin Gulch, Bob Moeller, Port aux Port Peninsula, Mae and Harry Becker, Bull Canyon Gorge, Valentine Zetlin, The Narrows, Larry Brown, Antelope Springs, Walt Boynton, Boquillas, Larry Davis, Amphitheater Lake, Lyman Guptil, 4-Hole Swamp, Johnny Fisk, Canyonlands, Bob Binnewies, Dougherty Point, Grace, Tim and Dan LeFever, Sharon Audubon Center, Jim Swan, Quoddy Head, Bruce Hawkins, Mamou, Hopi friends Mike, Terrance and Abbot, Steven's Arch, Rachel Bereson, Death Hollow, Penny, John, Sonya and Rufus, Navaho Mountain, Rhoda and Merril Abeshaus, Yosemite National Park, Nancy Neilsen, Gros Morne, Larry Becker, Kissimmee Prairie, Jan Beyea, Otter Creek, Jim and Loma Griffith, Sage's Ravine, Danny Lopez, Olympic Rain Forest, Doc and Marion Hodgins, Okeefenokee Swamp, Brian Bedell, Moosehorn Wildlife Refuge, Otis Sawyer, Wheeler Peak, Bill Bonyon, Eldon's Cave, Herbert and Maria West, Awatovi, Harry Vose, Pine Islands, Gary Hirshberg, Congaree River, Paul Howard, Music Canyon, John and Sue Tishman, Green Dragon, Doug Chapman, Amonoosuc Ravine, Joe Hickerson, South Rim Trail, Bob Voight, Race Point, Pete Seeger, Zuni, Dori Ells, Superstition Wilderness, Arnold Davis, Mary Huegel, Russ Peterson, Smuggler's Notch, Paul Lauzier, Indiana Dunes, Sylvia Cooke, Head Harbor, Bernie Yokel, Cape Hatteras, Annette Liberson, Shark Valley, Nate Levine, Taylor's Slough, Ralph Lusich, Martha's Vineyard, Charles Fitzgerald, Josephine's Sedge, Dick Wylie, Hurricane Pass, Markham Breen, Joshua Tree, Sheryl Crockett, Toshi Seeger, Esther and Art Wilson, Eddie Williford, Nickwackett Bat Cave, Mel, Bob, Jesse Blanchard, Devil's Kitchen.

Appendix F

When we came home, the sea said, "Wait".
We waited.
That's the way it is here;
sometimes the wind says, "Shiver",
and we shiver;
sometimes I make a shopping list,
but if the snow says, "Stay",
we clean the cupboard and make soup.
When we came home, and found the tide
over the road and into the alders,
I didn't mind. The water rose quietly,
and when it ebbed and we passed,
we went quietly.
That's the way it is here, sometimes,
and then we remember that it is no small thing,
to live in a holy place.